Solar System Voyage

In the last few decades, the exploration of our Solar System has revealed fascinating details about the worlds that lie beyond our Earth. This lavishly illustrated book invites the reader on a journey through the Solar System. It starts by locating our planetary system in the universe, then describes the Sun and its planets, the large satellites, the asteroids and the comets. With photographs and information from the latest space missions, readers will discover the lunar plains scarred by asteroid impacts, the frozen deserts of Mars and Europa, the continuously erupting volcanoes of Io and the giant geyers of Triton; they will cross the rings of Saturn, plunge into the clouds of Venus and Titan, and survive the spectacular crash of the comet Shoemaker–Levy into Jupiter, to emerge with a greater appreciation of the hospitable planet we call home.

Serge Brunier

Solar System Voyage

Translated by Storm Dunlop

CAMBRIDGE
UNIVERSITY PRESS

PUBLISHED BY THE PRESS SYNDICATE OF THE UNIVERSITY OF CAMBRIDGE
The Pitt Building, Trumpington Street, Cambridge, United Kingdom

CAMBRIDGE UNIVERSITY PRESS
The Edinburgh Building, Cambridge, CB2 2RU, UK
40 West 20th Street, New York, NY 10011–4211, USA
477 Williamstown Road, Port Melbourne, VIC 3207, Australia
Ruiz de Alarcón 13, 28014 Madrid, Spain
Dock House, The Waterfront, Cape Town 8001, South Africa

http://www.cambridge.org

French editions © Bordas/H.E.R., 1993, 1997, 2000
English translation © Cambridge University Press 2002

First published in French as *Voyage dans le Systeme solaire*, by Serge Brunier 1993
First English publication 2002

Printed in Italy by G.Canale C.& S.p.a. – Borgaro T.se – Torino – Italy

A catalogue record for this book is available from the British Library

ISBN 0 521 80724 7 hardback

CONTENTS

Introduction: the story of the planets 8

A star lost in infinite space 14

Our Star: the Sun 24

Mercury: baked by the heat of the Sun 32

Venus: a vision of Hell 38

The Earth: the story of a living planet 48

The Moon: setting foot on another world 68

Mars: a trip to the desert planet 82

Phobos and Deimos: pebbles in the sky 104

Gaspra: our first asteroid 108

Jupiter: the planet of storms 116

Shoemaker–Levy: timetable to collision 128

Io: the Volcano planet 134

Europa, Ganymede, Callisto: terrae incognitae to explore 142

Saturn: the Lord of the Rings 150

Titan: an Earth in hibernation 160

Enceladus and the worlds of ice 166

Uranus: a recumbent giant 172

Halley: the great traveller 178

Neptune: the great blue sea 192

Triton: volcanoes of ice 202

Pluto and Charon: planets in limbo 208

Appendices

Positions and motions in the Solar System 214

Telescopes and spaceprobes 226

Amateur observation of the planets 235

Bibliography and recommended reading 244

Index 245

■ EVER SINCE ANCIENT TIMES, A
SOLAR ECLIPSE HAS STRUCK A CHORD
WITH HUMAN IMAGINATION, WHICH
SAW IT AS A MANIFESTATION OF THE
DIVINE. THIS CELESTIAL
PHENOMENON, WHICH IS,
WITHOUT DOUBT, THE MOST
BEAUTIFUL SIGHT NATURE
AFFORDS, OCCURS WHEN THE
SUN, THE MOON, AND THE
EARTH ARE PERFECTLY ALIGNED.
THIS IS THE ECLIPSE OF
1991, PHOTOGRAPHED
FROM THE VOLCANO OF
MAUNA KEA ON THE
ISLAND OF
HAWAII.

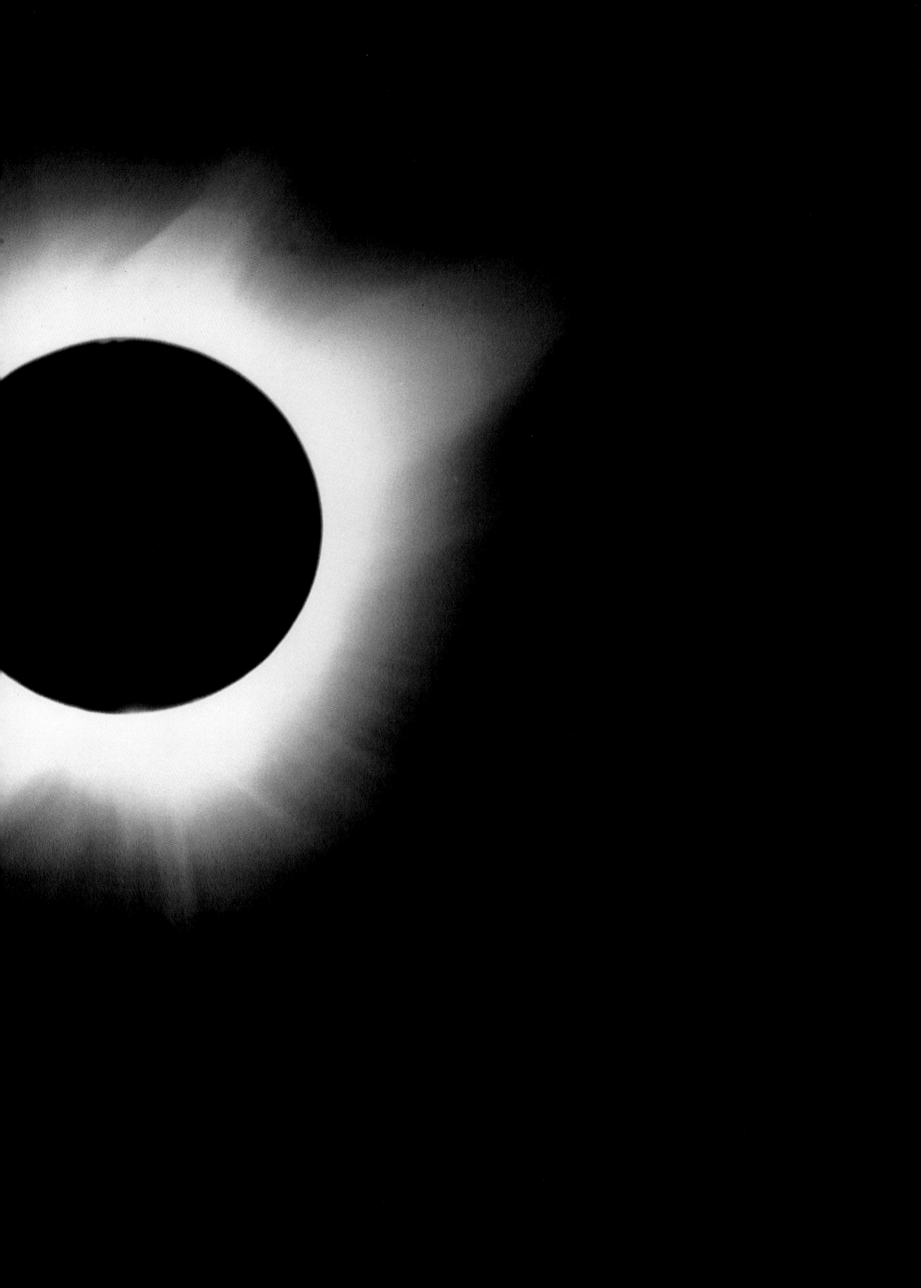

THE STORY OF THE PLANETS

In ancient times, there were five of them, which, with the Sun and the Moon, passed through the constellations of the zodiac in a seemingly erratic and incomprehensible fashion. The Greeks called them *planetos:* the wanderers. It took astronomers more than two thousand years of observations and calculations to understand the complicated workings of the giant celestial clock that controlled the whole world, influencing the fate of men and regulating the cycle of the seasons.

To the ancients, the Earth was the centre of the universe, and the Sun, Moon, Mercury, Venus, Mars, Jupiter, and Saturn revolved around it, embedded in crystalline spheres, whose motions were of greater or lesser complexity. Beyond them lay the *Sphere of the Fixed Stars:* the star-studded vault itself. As seen by the naked eye, the planets were no more than simple points of light in the sky of differing brightness and colour. People saw them as representing the characters of their individual gods: this one was moody, that one martial, and that one genial.

The arrival of a heliocentric view, which had been hovering in the wings since the time of the Greeks, but which had been held back for many, many years for theological reasons, sounded the death-knell for the concept of the heavens as being magically imbued, perfect, eternal, and divine. The new paradigm, by dethroning the Earth from its central, privileged position, endowed it with the same spatial rank as the Moon and the planets.

Who changed the status of planets in our society in this way? Was it Copernicus, who, in his posthumous work, deprived the Earth of its position at the centre of the universe, and bequeathed us our current concept of the Solar System? Was it Galileo, who was the first to observe the planetary system as a whole by means of his telescope, and who realised that, like the Earth, the brightest bodies in the sky were actually *other worlds?* Was it Neil Armstrong, who was the first to set foot on the surface of the closest of these worlds? There can be no doubt that such a profound change in our world-view did not take place overnight, and neither the publications of Copernicus, nor of Galileo, nor the televised broadcast of 'a small step for man' were sufficient in themselves to overturn our view of the world.

The evolution of ideas is not a single, monolithic process, and actually advances as a series of steps, of greater or lesser size. Although Copernicus and Galileo influenced their own times, and radically transformed our perception of the universe, initially their ideas reached a tiny fringe of scholars, philosophers, politicians,

■ Ever since people have raised their eyes to the star-studded sky, the night-time spectacle has remained unchanging. Our view of the world, by contrast, has evolved greatly, and henceforth we will always be able to see a cloud, planets, nearby bright stars, and the sparkling, distant arc of one of the Galaxy's spiral arms in their true perspective.

and prelates, before slowly diffusing through deeper and deeper layers of society. But who, in the 17th century, really worried whether it was the Earth or the Sun that revolved around the other?

In the majority of contemporary human societies, the status of the Earth and of the planets will, henceforth, be perfectly clear. They are no longer mysterious in nature; no longer offer scope for philosophical or religious speculations. The latter role has devolved upon the cosmology of the Big Bang, the point that we believe to be the origin of the whole of the universe, and the one point at which all current metaphysical questions converge. Admittedly, astrologers, who see signs in the sky and believe in predestination, still bear witness to the human race's intellectual inertia, and to the influence of the ancient paradigm, but then their art, which is becoming increasingly symbolic, is undoubtedly moving away from the 'true heaven of the stars' that the philosopher Alain evoked, and is totally divorced from the science of astronomy. Now that the year 2000 has passed, and when the irrational millennial angst that gripped some of our number has subsided, its current

vogue will decline, before, perhaps, reappearing later, under other auspices and in new forms.

The planets are other worlds. We owe this almost innate knowledge of our planetary environment, which is common knowledge particularly among children today, not to a 16th-century Polish canon, nor to a Florentine Renaissance scholar, and still less to a 20th-century test pilot, but to a large community of robots, which, in the space of just two generations, have visited the whole of the Solar System. The product of western science and technology at the end of the second millennium, these probes, bristling with communications antennae and electronic cameras, have been controlled from Earth. Thanks to them, or rather thanks to the researchers and engineers who created them, the deserts of Mars and the rings of Saturn are as familiar to us today as the most awe-inspiring landscapes of our own planet. The latter, oddly shrunken in recent decades by communications satellites and the growth of the civil aviation industry, contains hardly any exotic lands today. And although the last few adventurers still inspire us to dream, with their tales of crossings of the Sahara, the Himalayas, or Antarctica, we are only too aware that the blue planet no longer offers any real prospect for exploration. Adventure lies elsewhere: on the Moon and on Mars, and later, perhaps, on Io, Titan, and Triton.

Whilst we wait for humanity to return to the Moon, at the beginning of the third millennium, and for it to set foot on Mars – probably within thirty years – giant metallic insects are scuttling in every direction across the Solar System, allowing us to journey by proxy. And what a journey it is! Without the Venera and Magellan probes, we would know nothing of the strange, infernal world that lies hidden beneath the clouds of Venus. Without Mariner, Viking, Mars Pathfinder, and now Mars Odyssey or Mars Express we would be unable to prepare for our own future journey to the red planet. In some thirty years, these spaceprobes have brought greater progress to our knowledge of the Solar System than two thousand years of astronomical observation.

This first wave of Solar-System exploration is not without an echo of the great maritime expeditions of the past. Like Magellan, La Pérouse, or Darwin, our space-probes depart for years, or even decades, at a time, occasionally becoming lost on the way, swallowed up by the darkness of space. Others, every now and then send us back their regards. They are so old that two genera-tions of researchers have succeeded one another in looking after them. The Pioneer 10 spaceprobe, con-ceived in the 1960s, and which left Earth in 1972, has subsequently left the Solar System, and is some ten billion kilometres distant, but still remains in radio contact with our planet. Some of the engineers who listen to it today with the giant Goldstone radio dish in the Mojave Desert, are as old as the spacecraft itself.

Like the great voyages by the naturalists of the 18th and 19th centuries, those carried out by these spaceprobes have taught us a lot, and have sometimes completely altered our view of the world. Two great lessons have emerged from our exploration of the Solar System. The first was expected. Well before the first rocket left the Earth, researchers intuitively understood it. It is a law of nature, and the very basis of science. It tells us that throughout the local universe, the same causes produce the same effects. The laws of physics are the same on Earth, Mars or Neptune. Meteorological and geological phenomena are not the prerogative of our own blue planet. Fogs, clouds, glaciation, the cycle of the seasons, aerial erosion, and volcanism are universal phenomena.

By contrast, the second lesson that exploration of the Solar System has taught us was a surprise. It showed us that the range of ways in which these grand planetary laws are applied is almost infinite. Henceforth, diversity and complexity will be the key-words that researchers apply in describing the bodies in the Solar System. Because, contrary to what astronomers thought even some twenty years ago, the planets and satellites do not fall into simple categories, despite having similar physical and chemical characteristics. For example, the classic idea that bodies with a low mass do not vary greatly from one another, remaining fixed in the state in which they were formed 4.5 billion years ago, has proved to be completely false. The smallest bodies prove to be the most complex, and exhibit the most varied types of landscape. Io, Europa, Enceladus, Titan, Miranda, and Triton have defied all calculations, and their origi-nality and exotic nature have surpassed the specialists' most outrageous models.

The astronomers' surprise is perhaps partially explained by their scientific culture, which was extremely theo-retical and still deeply affected by two centuries of unremitting study of celestial mechanics. To many researchers, in fact, the planets were formerly nothing more than individual bodies moving around the Sun. The change in status of the planets – which became worlds in just a few years – surprised them in just the same way as people in the street, lifting their eyes to the night sky on a summer evening in 1969, suddenly became aware that two of their fellow creatures were walking on the surface on that shining disk that resembled nothing so much as a sad, white-faced clown.

Nowadays we perceive the planets differently. To understand them, we need to study them both in space and in time. Their current appearance bears witness to their past evolution, the conditions that held sway in their vicinity 1, 2, 3, or even 4 billion years ago, but also to the subtle interactions that they have undergone with other bodies in the Solar System.

In the final analysis, 4.5 billion years of evolution have not reduced the surfaces of the planets to a single level, minimized geological differences, nor produced iden-tical landscapes everywhere. On the contrary: everybody

MANKIND HAS LEFT THE
EARTH. FOR SOME TWENTY
YEARS, TERRESTRIAL ORBIT HAS
BEEN PERMANENTLY OCCUPIED
BY ASTRONAUTS, WHO HAVE
SEEN OUR PLANET WITH NEW
EYES. THEY ALL DREAM OF THE
DAY WHEN OTHER COSMIC
EXPLORERS WILL RETURN TO
THE MOON AND VISIT
MARS, BEFORE SETTLING
THERE PERMANENTLY TO
CREATE A NEW
CIVILIZATION.

in the Solar System is different. Although Jupiter and Saturn have experienced convergent evolution and have come to resemble one another like sister planets, they have, on the other hand, given birth to a wonderfully diverse range of satellites. Although Venus and Earth originally had the same physical, chemical and geometric characteristics, their respective fates subsequently diverged so radically that the two bodies are today so dissimilar that, strictly speaking, they no longer have anything in common.

■ THE MOON AND VENUS IN CLOSE CONJUNCTION AT SUNSET. AS SEEN FROM EARTH, THE MOON AND THE PLANETS OCCASIONALLY MEET IN THE SKY. THESE APPARENT CLOSE APPROACHES ARE SIMPLY THE EFFECT OF PERSPECTIVE, BECAUSE THE TRUE DISTANCE BETWEEN THESE BODIES IS RECKONED IN TENS OF MILLIONS OF KILOMETRES.

Just like living species, planets are subject to chance, and are subject to a greater or lesser degree of restraint from environmental pressures. Despite having a common origin, their history is unpredictable. Run the film of the evolution of the Solar System, once, one hundred times, or a thousand times, and each time the story will vary from the preceding version. The true lesson from our venture into space is perhaps this: Although the largest structures in the universe – clusters, galaxies, nebulae, and stars – are repeated in identical, or almost identical fashion, the smallest and the most complex structures in the universe (the planets) are unique. Without prejudging the existence of other planetary systems in the universe, we can assert without fear of contradiction that there exists only one Solar System.

And what should we say of the Earth? Our world is, in some respects, the duck-billed platypus of our planetary system. In fact, this deep blue planet seems, somewhat arrogantly, to be host to all of the strangest peculiarities

in the Solar System. First, together with the Moon, it forms a pair of planetary bodies, with unique dynamical properties. Second, we find, randomly across its surface extremely rare, or even unique features: a thick atmosphere that produces a greenhouse effect, powerful volcanism, intense plate tectonics, and water existing in all its three basic phases. It is probably the random and heterogeneous juxtaposition of these characteristics that allowed the unbridled expression of a physical and chemical activity – namely, life – to appear, as well as what is, above all, the most extraordinary of nature's manifestations: ourselves, and *self-awareness*.

Although science may have removed the magic and the divinity from the planets in the Solar System, and although it has eliminated all the mystery behind the perpetual motion of the heavens, it has, on the other hand, meant that these worlds have become more real, more vibrant, but also more unpredictable, and thus more fragile. The Solar System has its own history, which reaches back to its origin 4.5 billion years ago. Each year we understand a little better the successive episodes in its long and turbulent evolution. We also know that the Solar System has long since reached maturity, and that, like living organisms, will eventually come to an end. This will take place quite suddenly, in about five billion years, with the transformation of the Sun into a red giant, and the subsequent ejection of its outer layers into space.

Is this to say that the future of our planetary system is already decided? Are we already able to predict the evolution of the bodies of which it is formed? Following in the footsteps of August Comte, positivist philosophers and astronomers thought so for many years. But chaos theory, which was born through the intuition of Henri Poincaré around the turn of the 19th and 20th centuries, and which has been applied to the physical sciences since the 1960s, put an end to this school of thought, and killed off this proud and foolish hope. Chaos theory is a complex and fertile mathematical tool, which could not be developed and become established until the appearance of extremely powerful computers. It is used to study the stability over time of complex and periodic physical systems, such as that found in the Earth's atmospheric circulation, or the orbital motion of stars in the Galaxy.

When applied to celestial mechanics – which, historically, was regarded as the most deterministic of all sciences – it allows us to see that, over extremely long time-scales, the future state of the Solar System cannot be predicted. The classic image of the heavens as unchanging, with perfectly regular motions, a legacy of the Greeks, must be abandoned – and this time for good. Like meteorologists, who no longer try to predict what the weather will be a week in advance, astronomers henceforth have to abandon any thought of calculating the positions of Mars or Pluto one hundred million years from now.

The philosophical view of the universe provided by modern physics is astounding. The last century has seen the birth of quantum mechanics, which also predicts that there is a fundamental limit to our knowledge of the microscopic world. The general acceptance that the cosmos is expanding has given the universe as a whole a historical, evolutionary, and unpredictable status, which it never had previously in the whole of the history of science. At long last, variable and indeterminate phenomena also appear to govern the behaviour of the planets – the very last of the astronomical bodies that were thought to be immutable. Two thousand years after the Greek Miracle, the Renaissance, in the form of Giordano Bruno, Copernicus, and Galileo, freed mankind from too close a relationship with nature and with God. Today, science goes even further, by freeing the heavens themselves from the rigid shackles of the laws of nature. Hidden behind the horizon of predictability imposed by chaos theory, the future of the Solar System – our future – is henceforward wide open.

A star lost

in infinite space

■ THE INTERGALACTIC LANDSCAPE AS PHOTOGRAPHED
FROM THE EARTH, SOME 28 000 LIGHT-YEARS FROM THE
CENTRE OF THE GALAXY. BEHIND THE UNIFORM SCATTERING
OF FOREGROUND STARS, A GLOWING NEBULA CASTS ITS
GASEOUS DRAPERIES AGAINST THE VELVETY BLACKNESS
OF THE SKY. MUCH FARTHER AWAY, AT A DISTANCE OF
SEVERAL TENS OF MILLIONS OF LIGHT YEARS, THERE
IS THE DELICATE VISION OF A SPIRAL GALAXY,
APPEARING LIKE A SILVER SPINDLE AGAINST
THE BACKGROUND SKY.

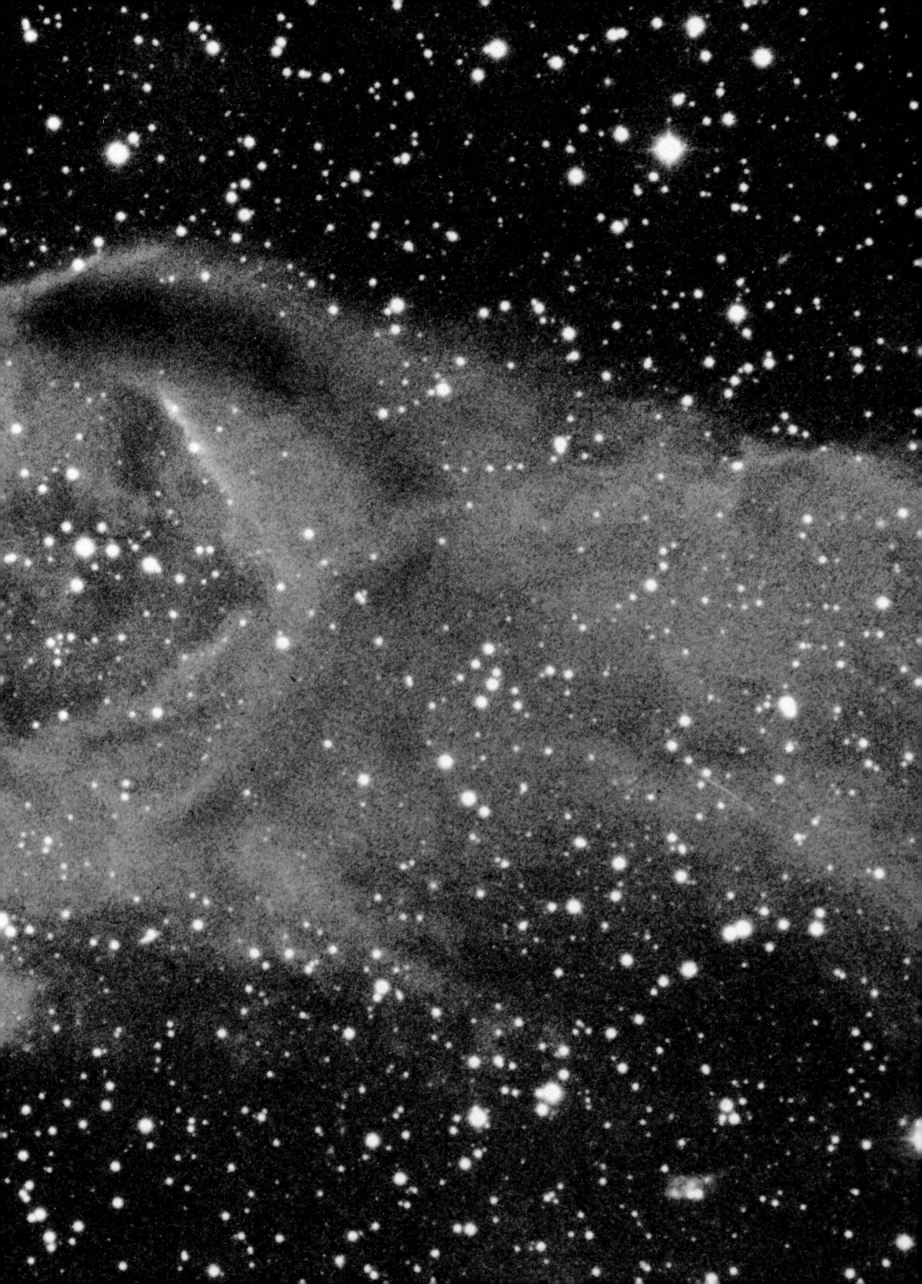

A BREATHTAKING COSMIC PERSPECTIVE: A BRILLIANT NEARBY STAR IN THE MILKY WAY BLAZES IN FRONT OF A DISTANT GALAXY. WITHIN GALAXIES STARS ARE BORN, EVOLVE, AND FINALLY DIE.

It looks like a living thing. Winding, infinitely slowly, its silvery arms around its hot, brilliant centre, the Galaxy floats in space. It might well be a large, solitary snowflake, lost in empty, frigid space. You feel that you would like to cup it in your hands, capture its soft glow, and stroke it. You almost fear that it might melt or break. But this fragility is an illusion, because this wondrous heavenly snowflake, this crystal, which has been rotating, quite indifferent to any emotions that it arouses, for more than 10 billion years, is a universe in itself, which has seen thousands of millions of worlds born, and die, within it. Because this flake, this galaxy, actually consists of stars. A multitude, a myriad, of stars. They are so numerous that millions of them seem to merge into diaphanous clouds, so many that no one will ever be able to count them, and so many that even their number has no real significance. And, lost within these hundreds of thousands of millions of its sister stars – we do not know exactly where – gently shines our own star: the Sun.

The Galaxy is one of the building blocks of the universe. Astronomers on Planet Earth continue to call it the Milky Way, as a relic of Antiquity, when it was thought to have gushed from the breast of Juno. From any of the many possible planets that it may shelter, it will always appear in the same guise: veils of nebulae and clouds of stars, arching over the celestial sphere; a grand, all-embracing silhouette, reassuring and always present. The Milky Way is an omnipresent element in a planetary landscape.

The universe, to the greatest distances that we are able to probe it, both in space and time, is filled with galaxies just like our own. All similarly consist of stars; all have had the same histories, and all have evolved in the same way until our day. Galaxies tell us the history of the universe as a whole. Cosmology teaches us that they were all born some twelve to fifteen billion years ago, from a dense, hot soup of expanding matter, just as lumps appear in cooling cream. This hot, expanding gas itself arose by expansion from a single point in space and time, whose properties are currently unknown, which was literally the origin of the universe: the Big Bang. Modern-day physics has not been able to model this primordial explosion, nor discern its cause – if it had one. Despite piercing ever closer to the – asymptotic – origin of the universe, our telescopes provide us with an image that is

FAR BEYOND A SCATTERING OF STARS IN THE CONSTELLATION OF HYDRA, A SPIRAL GALAXY FLOATS IN THE ICY VOID OF SPACE. SEEN FROM A DISTANCE, OUR OWN, THE MILKY WAY, PROBABLY RESEMBLES THIS FRAGILE JEWEL. LIKE IT, THIS GALAXY INCLUDES MORE THAN 100 BILLION STARS, WHICH FORM DELICATE SPIRAL ARMS, THE RESULT OF A SLOW, STATELY ROTATION.

still vague and indecipherable. The universe remains resistant to all our tools, whether they be technological or conceptual ones, and astrophysicists, philosophers, and theologians will perhaps discuss the nature of existence until the end of time. But does the universe really have any meaning?

Quite apart from the Big Bang – which still remains nothing more than a theory, despite being almost unanimously accepted – whole swathes of the universe's intelligible history remain in the dark. How did the four forces of nature – the strong nuclear force, the weak nuclear force, electromagnetism, and gravity – which govern the world and enable us to understand it, actually appear? Why and how did fluctuations occur in the expanding cosmic plasma? These questions may perhaps be answered during the course of the current century. The evolutionary path that matter has traced from the primordial explosion is, itself, quite well known. For some 15 billion years the universe – space-time and energy/matter – has thus been in constant expansion. From time zero, with this expansion, its spatial dimensions have never stopped increasing, while its density and temperature have de-

■ It is impossible to gain an idea of the real appearance of our Galaxy, because our planet lies inside its great spiral disk. Seen from Earth, the Galaxy appears as an immense milky arch spanning the celestial vault from one side to the other. Its central bulge may be seen in the direction of Sagittarius.

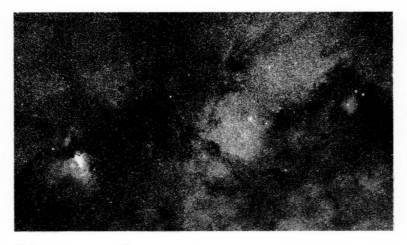

■ Paradoxically, our Galaxy is one of the few where we cannot see the centre. Because the Solar System lies precisely in the galactic plane, our line of sight towards the centre, some 28 000 light-years long, is totally blocked by millions of stars and clouds of interstellar gas and dust.

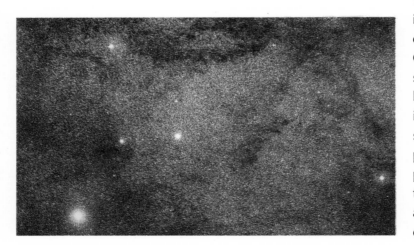

■ Observed through a telescope, the region around the centre of the Galaxy is hidden by a curtain of stars. The millions of individual objects look so close that they appear to touch one another. In reality, there are thousands of millions of kilometres between them. Despite appearances, the Galaxy is essentially empty.

been compensated by a tendency towards ever greater organisation of matter within galaxies. Hospitable refuges in a space that was ever becoming more empty and cold, galaxies started to produce stars.

Like venerable alchemists, the latter have, over the course of the aeons, patiently transmuted lead into gold. From their nuclear crucibles have burst forth carbon and sulphur, iron and silicon, oxygen and nitrogen. Over those 15 billion years, innumerable generations of stars have followed one another, each enriching the interstellar medium with the atoms they have created. The Galaxy has stirred it all together, being lit up every year by twenty-odd new stars, and paling every hundred years in the glare of a supernova. This stellar nucleosynthesis is the origin of all the known elements. With time, the Galaxy, which originally consisted of just hydrogen and helium, has clothed itself in curtains of oxygen, and sparkled with gold and platinum. Without the slow processes of stellar evolution, our Sun, the planets, and the Earth itself would not exist. Like them, we ourselves have been created from the fertile ashes of ancient generations of stars.

The Galaxy is gigantic: to

creased. Radiation and elementary particles appeared first, as did, at the end of one million years, the very first atoms, hydrogen and helium: these are the primordial gases. It is assumed that these gases subsequently condensed into galaxies, one billion years after the initial explosion. Since then, still constantly expanding, the universe has continued to cool and become more empty, but this headlong rush to oblivion has

cross its lens-shaped disk, light, travelling at a constant 300 000 km/s, takes nearly 80 000 years. It therefore measures 80 000 light-years across, to use one of the units of distance employed by astronomers. Translating this into a more human unit, the kilometre, does not make much sense: it is a figure of 800 000 000 000 000 000 km. The only significant meaning of this dimension is that it forbids us, undoubtedly for ever, from

travelling across the Galaxy and, even more importantly, escaping from it. We shall never know what it really looks like, and shall be forever condemned to dream of the incredible spectacle that our own galaxy presents, merely observing from a great distance the thousands of millions of its counterparts that are scattered throughout the universe.

Although we do not know how many stars make up the Galaxy, thanks to studies of its overall mass and its motion, we are able to roughly estimate its tally as 200 billion stars. In amongst this almost infinite multitude of stars, our own Sun does not stand out. A medium-sized, medium luminosity star, and inevitably used as a standard of reference by astrophysicists, it is accompanied on its galactic rotation by stars that are one tenth of the size, and which are one millionth of its brightness. These pale red dwarfs have an almost infinite life expectancy: 50–100 billion years. The fact is that no stars of such a great age apparently exist anywhere in the universe – on the contrary, the very oldest are 15 billion years old. This is considered by astrophysicists as reliable proof of the accuracy of the Big-Bang theory. Alongside this veritable army of invisible, or nearly invisible red dwarfs blaze stars that are true cosmic lighthouses; stars so rare and so bright that, visible from one side of the Galaxy to the other, they could well serve as interstellar beacons, if there were anyone who could travel through space. Deneb, Rigel, Saiph, Canopus, Bellatrix, and Polaris (the Pole Star) each blaze forth like nearly 100 000 Suns!

The 200 billion stars in the Galaxy slowly revolve around its centre, their speed decreasing, the farther they are from that centre. The Sun, at a distance of some 28 000 light-years from the Galaxy's rotation axis, takes 250 million years to complete one revolution. As seen from Earth, the galactic disk looks very dense, with the stars crammed up against one another. This is an illusion caused by perspective: the average distance between the stars is about 3 light-years, i.e., 30 million million kilometres! In an attempt to obtain some mental image of the Galaxy, we need to make an experiment. Imagine that every star in the Milky Way is represented by a grain of sand. On this scale, the Galaxy would be a disk 300 000 km in diameter! Imagine the volume represented by a child's sandpit spread over the Earth–Moon distance: the Galaxy essentially consists of empty space.

On the galactic scale, the Sun is a young star. Although, speaking in round figures, the Galaxy and its first generation of stars was born around 15 billion years ago, our own star is different. Consisting of gas created and then expelled by the first generations of stars, our own is rich in heavy elements, such as carbon, oxygen, and nitrogen. Our star was born approximately 4.55 billion years ago, in the dense cocoon of a nebula. Like all stars, the Sun formed as a result of the gravita-

tional collapse of a cloud of gas. Deep within one of the nebulae that form repea-

tedly in space each time one of the Galaxy's spiral arms passes by, a globule of denser gas slowly collapsed in upon itself. In so doing, it naturally assumed the form of a rotating disk, at the centre of which a sphere of gas started to collapse inwards, becoming hotter and denser as it did so. The unequal battle between the force of gravitation – which tended to cause the matter to fall towards the centre – and the gas pressure, ended with gravity gaining the upper hand, and the sudden ignition of the nuclear fusion reactions at the star's core: the Sun had been born.

It was violent birth: the young star was unstable and turbulent, and blasted into space the remnants of the nebula that gave it birth; it shot out incredibly energetic jets of matter, whilst around it the disk of gas and dust became organised and formed the Solar System. Almost perfectly flat, this disk was swept by the solar wind, a particle plasma arising in the star's searing outer atmosphere. The lightest elements, such as hydrogen and helium, were blown out towards the edge, while the heavy atoms remained evenly spread throughout the protoplanetary disk. The most dense particles, consisting

of carbon and silicate dust, clumped together, creating – over the course of thousands of years – embryonic planets: large, irregular, and fragile lumps of material. Sweeping space with an ever-increasing efficiency (because their cross-section increased as the accretion continued), they literally cleaned up the Solar System. This was a long, complex process: every object that was forming was prone to be partially or completely destroyed by collision with another planetoid, thus feeding dust back into surrounding space.

Nevertheless, eventually, safe zones developed at increasing distances from the Sun, where the modern planets finally formed. Each object, depending on its mass, either captured the interplanetary material, or expelled it towards the outer edge of the Solar System. Close to the Sun, were the small, dense planets of Mercury, Venus, Earth, and Mars. Farther out, there were the giants, Jupiter, Saturn, Uranus, and Neptune. It was an uncertain, dangerous period. Mercury, hit full force by an asteroid, was nearly destroyed. When they entered forbidden zones, where the gravitational gradients were too strong, certain planetoids fragmented, creating magnificent rings around the giant planets. Over the millennia, the struc-

ture of the Solar System gradually became established. The planets captured smaller bodies, surrounding themselves with satellites, interplanetary space was swept clean, collisions became rarer, and the Solar System took on the peaceful appearance that it has today.

Farther out in the Solar System, the scenario was exactly the same, even though the composition of the planets was different. The four giant planets are principally gaseous, because they captured some of the gas contained in the original nebula. As for the small, distant bodies, such as Pluto, Triton, and Titan, they largely consist of light elements – water, nitrogen, methane, etc. Nevertheless, even 5 billion kilometres from the Sun, these bodies are covered in impact craters, identical to those that pock-mark the Moon and Mercury. The Solar System was forged beneath an incessant bombardment by billions of asteroids.

To classical astronomy, before the space age began, the Solar System consisted of nine planets. In order of distance from the Sun, these were: Mercury, Venus, Earth, Mars, Jupiter, Saturn, Uranus, Neptune, and Pluto. Orbiting these planets were some fifty-odd satellites, ranging from 10 to 5280 km in diameter, such as Atlas, Phobos, the Moon, and Ganymede. In fact, based on their masses, their sizes, and

their physical and chemical characteristics, many of these satellites
resembled planets in their own right,
and some were even larger than what
were classically regarded as planets. It
has become acceptable to use both
the terms planet and satellite when
describing these objects, and this is
what we will do in this book. By
contrast, asteroids and comets, which
populate the Solar System in their
thousands, are very different from the
planets. Generally tiny, they have too
little mass to be moulded by the force

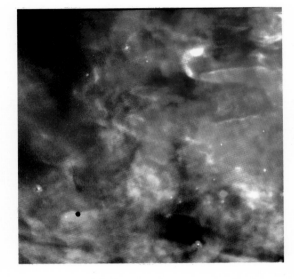

■ THE COMPLEX GASEOUS STRUCTURE OF THE FAMOUS NEBULA,
M42, IN THE CONSTELLATION OF ORION, IS HERE CAPTURED BY
THE HUBBLE SPACE TELESCOPE. THIS REGION IS VERY YOUNG,
AND OVER SEVERAL THOUSAND YEARS, HUNDREDS OF STARS HAVE
BEEN BORN HERE.

of gravity, like a true planet. They are irregular in shape, rather
than spherical, and they are undifferentiated, having been
born without a molten core, a mantle, or a crust.
Planetologists consider that the sole intact witnesses of the
birth of the Solar System are these large rocks, which have
been wandering, menacingly, between the planets for 4.5
billion years.

As seen from Sirius, the planets in the Solar System are
insignificant: all that exists is the Sun, whose brilliant light may
be detected at distances of tens of light-years. In fact, when it
was formed the Sun cornered 99.9 per cent of the total mass of
the Solar System! A thousand times as massive as Jupiter, the
largest of the planets, it is some 333 000 times as massive as
the Earth. Like the Milky Way, the Solar System is essentially
empty. At its centre, the Sun is 1.4 million kilometres in
diameter. The first planet, Mercury, lies at a distance of
58 million kilometres, Neptune and Pluto, the outermost, at
4500 million kilometres. If one were to represent the Sun by
an orange, the Earth would be just a tiny ball, 1 mm across,
located 11 metres away! Saturn would
be a ball, 1 cm across, at 100 metres;
and Pluto would be found at about 600
metres distance. All the planets
revolve around the Sun in the same
direction, practically in the same
plane, and in almost perfectly circular
orbits. Being subject to the universal
law of gravitation, the closer they are
to the Sun, the greater the attraction
they experience. They counteract the
Sun's gravitational field, which tends
to pull them towards it, by their own
individual orbital motion, which tends
to cause them to recede. Indeed, their
orbital velocity decreases as a function
of their distance from the Sun. Mercury
takes just 88 days to complete an

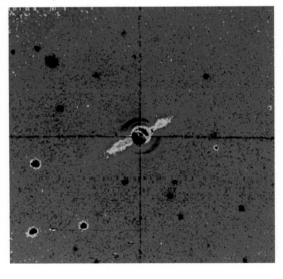

■ THE DISK OF DUST THAT SURROUNDS THE YOUNG STAR BETA
PICTORIS, WHICH LIES AT A DISTANCE OF 60 LIGHT-YEARS, IS
SLOWLY BEGINNING TO CONDENSE INTO PLANETS. ASTRONOMERS
HAVE RECENTLY DISCOVERED SUCH PROTOPLANETARY DISKS
THROUGHOUT THE MILKY WAY.

orbit, while Pluto completes its own –
admittedly 100 times as great – in
about 250 years! Mercury's orbital
velocity is 48 km/s, and Pluto's about
4.7 km/s.

The Sun is just one of the 200 billion
stars that populate the Galaxy. Its birth
was accompanied by that of a planetary system. Does this mean that other
stars may also be surrounded by a
retinue of planets? We now know that
the answer is 'Yes'. At this point in
time, the beginning of the 21st century,
astronomers already know of eighty-
odd newly discovered planets orbiting other stars. The very
first of these exoplanets was discovered at the end of 1995 by
two Swiss researchers working at the Observatoire de Haute-
Provence in France. Because it was impossible for them to
observe an exoplanet directly in the vicinity of its star – the
difference in luminosity of the two bodies is of the order of one
billion times! – Michel Mayor and Didier Queloz used an
indirect method of observation. This method, known as the
radial-velocity method, consists of detecting the evidence of a
slight movement of a star, backwards and forwards along the
line of sight, this irregularity being caused by the gravitational
perturbations of any possible planet. Since the historic discovery by Mayor and Queloz of the planet orbiting the star
51 Peg, the catalogue of extrasolar planets has been growing
by roughly one new planet every two months. The eighty-odd
planets discovered to date by American, Swiss, and Franco-
Swiss teams, almost all orbit solar-type stars. All these
exoplanets are giants, with a mass equal to that of Saturn or
Jupiter, or are sometimes two, four, or even six times as
massive as the largest of the planets in
the Solar System. Could the system
around 47 UMa in the constellation of
Ursa Major (the Great Bear), which
has two giant planets resembling
Jupiter and Saturn, harbour even more
planets, similar to Venus, Earth and
Mars? The planet known as 47 UMa b
orbits the star in about three years, at
a distance of 381 million km, while
47 UMa c takes slightly more than
seven years to complete one orbit, at a
distance of 560 million km.

Running in parallel with their search
for planets in the Galaxy, astronomers
are also keenly interested in the
method by which they are formed. To
understand how our Solar System was

born, and to compare its characteristics with those of other planetary systems, specialists are searching the Galaxy for young planetary systems that are captured in the actual process of formation. Several young stars have been found, such as Beta Pictoris and 55 Cancri, around which gas and dust have recently collapsed to form a disk, which will, in turn, collapse to form planets. As a result of all these investigations, it seems that the Solar System, with its nine planets, four of which are giants, is probably not at all exceptional. The system around the star Upsilon Andromedae, for example, has three giant planets, discovered between 1997 and 1999.

In the light of these discoveries made at the very end of the 20th century, astronomers are at last able to carry out a proper statistical study of the overall planetary population of the Galaxy. The result is staggering: according to the experts, there may be more than 10 billion planetary systems among the shimmering stars of the Milky Way.

When faced with this multiplicity of new worlds, astronomers are now wondering, more and more seriously, whether planets similar to Mars, Venus, or the Earth, do not exist elsewhere. Given that the laws of nature are the same throughout the universe, and that the same causes produce the same results, such a speculation is perfectly reasonable. All the stars that have been found to be accompanied by giant planets, such as 51 Pegasi, Upsilon Andromedae, 55 Cancri, 14 Herculis, Rho Coronae Borealis, Tau Boötis, 16 Cygni B, and their innumerable stellar brethren, similar to the Sun, are perhaps also accompanied by small planets like our own, but which we are, as yet, unable to detect. Undoubtedly the greatest scientific adventure of the 21st century will be to try to discover the Earth's twins elsewhere. Naturally, researchers have devised new instrumentation to ensure this search succeeds. The French Corot satellite, and the American Kepler satellite, should, from 2004 onward, be searching for other Earths by detecting the regular mini-eclipses they create as they pass in front of their parent stars. Darwin, a European project by the European Space Agency (ESA), may see the light of day between 2015 and 2020, is even more ambitious. This space interferometer will initially photograph tens of exoplanets, and then accurately analyze the composition of any possible atmospheres, in the hope of detecting proof of biological activity.

This brings us to the persistent and fundamental question: Are there other inhabited worlds in the Galaxy, other thinking beings who wonder, raising their faces or antennae to the sky, about the origin and the *raison d'être* of the universe? Some scientists think so, and have even been trying, for some fifty years, to listen to the stars with giant radio-telescopes, hoping to capture electromagnetic signals – whether intentional or not – emitted by another civilization. This has been a forlorn hope as yet, which doesn't really surprise most biologists. The history of evolution on Earth, and of the species that inhabit it, has, in fact, been an extraordinarily rich and complex skein, interwoven throughout by chance – and incapable of being reduced to the simple deterministic laws of physics. The emergence of a self-aware species is, in itself, an accident of galactic history. It is unlikely that the same scenario has been repeated elsewhere, so perhaps extraterrestrials do not exist.

It looks like a living thing. Winding, infinitely slowly, its silvery arms around its hot, brilliant centre, the Galaxy floats in space. Beyond it, beyond the sparse stars that form the limits of the Galaxy, is the unending void of the universe, punctuated at vast intervals by the glimmer of another galaxy, another island universe, shining with hundreds of thousands of millions of stars. The galaxies are innumerable, and astronomers ceased to try to count them when they realised that there were even more of them than the stars in our own Galaxy.

Our Star:

The Sun

■ A SOLAR PROMINENCE RISES ABOVE THE LIMB OF
OUR STAR. SUCH EVENTS OCCUR ALMOST DAILY ON THE
SURFACE OF THE SUN. PROMINENCES, WHICH EJECT
PLASMA INTO THE VACUUM OF SPACE AT SEVERAL
HUNDRED KILOMETRES A SECOND, MAY EXTEND FOR
HUNDREDS OF THOUSANDS OF KILOMETRES. THEY
ARE EPHEMERAL, EITHER RAPIDLY FALLING BACK
ONTO THE SUN, OR SLOWLY DISPERSING INTO
SPACE.

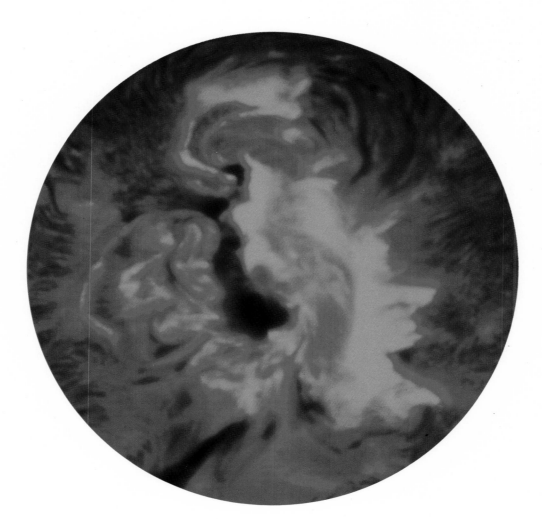

■ THE TEMPERATURE OF THE SURFACE OF THE SUN IS ABOUT 5500 °C. THE
STAR'S LINES OF MAGNETIC FORCE EMERGE FROM SUNSPOTS, WHICH APPEAR DARK,
BUT WHICH ARE, IN REALITY, STILL BLINDINGLY BRIGHT.

Scene: Somewhere in space, just a few million kilometres from the central star of a planetary system that is home to a living, intelligent species, which occupies the third planet, and which has begun to explore interplanetary space. This species has never risked its spacefaring robots in the neighbourhood of the star, which is the Solar System's dead zone. It is a desert, consisting of a vacuum and deadly radiation, that one can visit only in passing, buffeted by the powerful blast from the gigantic star, and vaporized by the energy that it sheds into space. Nothing can survive in close proximity to the Sun. From time to time, asteroids brush past the searing heat of the Sun, and escape, blackened, into calmer regions. Sometimes, comets approach a little too close, and fall into the body, are vaporized, or fragment.

The prominence slowly rises above the furnace-like surface of the Sun. Rising at nearly 1000 km/s, in a straight line, it seems to be capable of escaping from the Sun's incredible gravitational field. It is like a dream: the absolute silence of space removes all sense of the violence unleashed in the star. It is impossible to look at the surface of the Sun directly without being blinded. The deadly radiation coming from the

depths of the Sun are invisible, but would be fatal after just a few minutes of direct exposure. Bathed in this unbearable solar radiation, the great eruptive prominence, continuing to rise, is visible only at the limb of the star, where it is silhouetted against the almost completely black background of space.

After rising vertically for nearly a million kilometres, the jet of plasma, a delicate pink in colour, seems to slow down. The prominence starts to curve over, widens, and takes on the appearance of a fiery arch, that seems to hang suspended, stationary, high above the surface of the Sun. But then a ribbon of tenuous material detaches itself, and continues its path out into space, as the prominence finally sinks back towards the surface. The cloud of gas, having escaped from the Sun's gravity, will travel out through space, gradually becoming more tenuous, and vanishing over the course of a few hours. Some of its atoms will sweep past the desolate surface of Mercury, or will dive into the superheated atmosphere of Venus. In a few days, 150 million kilometres out, the Earth's skies will glow with the magnificent sight of aurorae, caused by the cascade of particles into the upper atmosphere, and these mauve and emerald-green curtains of light, which disappear with the

■ DESPITE ITS SEEMINGLY EXPLOSIVE AND UNPREDICTABLE APPEARANCE, THE SUN IS A REMARKABLY STABLE STAR. IT MAINTAINS ITS THERMODYNAMIC EQUILIBRIUM OVER THE COURSE OF TIME BY 'BURNING' FOUR MILLION TONNES OF HYDROGEN EVERY SECOND. AT THIS RATE, IT WILL REMAIN UNCHANGED FOR ANOTHER ONE TO TWO BILLION YEARS, BEFORE EVOLVING, VERY SLOWLY, TOWARDS THE RED-GIANT STAGE.

dawn, will mark the end of the solar eruption.

The Sun: one of the 200 or 300 billion stars that populate the Galaxy. Although average, our star is by no means uninteresting. Compared with the majority of stars in the Milky Way, the Sun is a rather massive, bright star. Most of the stars have one tenth of its mass, one tenth of its size, and have a tenth, hundredth, thousandth, or even a millionth of its luminosity. Only a few billion are more massive and brighter. The Sun was born 4.55 billion years ago, through the gravitational collapse of a cloud of

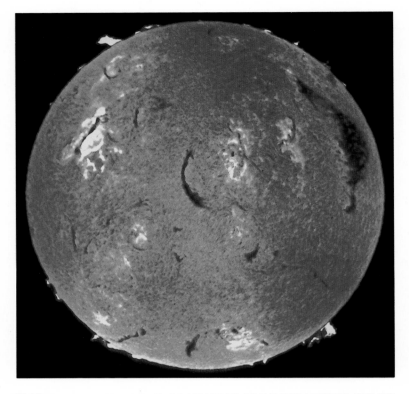

■ HERE, USING A SPECIFIC FILTER, WE ARE ABLE TO SEE THE CHROMOSPHERE OVER THE WHOLE OF THE SOLAR DISK. THE EXTREMELY HOT ACTIVE REGIONS ON THE SUN APPEAR IN LIGHT YELLOW, AS DO THE PROMINENCES THAT RISE AROUND THE EDGE OF THE DISK. THE DARK FEATURES VISIBLE IN THE CENTRE OF THE DISK ARE PROMINENCES SEEN AGAINST THE BRIGHTER SURFACE.

hydrogen and helium. When the gas pressure at the centre of the embryonic star reached a value – that is for us, utterly surreal – of 100 billion times the Earth's atmospheric pressure, the gas was at a temperature of over 10 million degrees Celsius. Under these extreme physical conditions, the particles have such high energies that they are able to fuse when they collide head-on. The fusion of four atoms of hydrogen produces one atom of helium, with a small fraction of the mass of the atoms being converted into radiation. Within the Sun's core, such collisions occur billions of times a second. For every second that passes, four million tonnes of hydrogen is completely converted into radiation! Our star is a gigantic thermonuclear reactor. With its output of 3.9×10^{26} W, the Sun creates more energy than one billion billion nuclear power stations. Obviously, on the Sun's scale, the four million tonnes of mass that are lost every second is negligible, and it is capable of continuing at this profli-

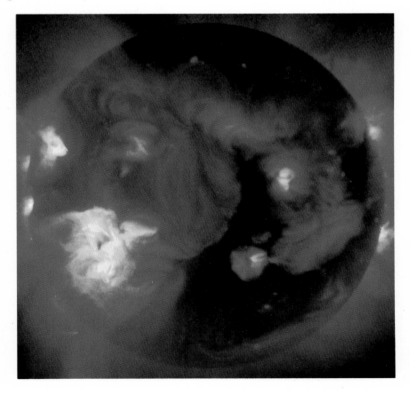

■ THE SOLAR CORONA, WHICH IS AN EXTREMELY TENUOUS AND EXCEPTIONALLY HOT MEDIUM, IS SEEN HERE BY A SATELLITE CARRYING AN X-RAY TELESCOPE. THE CORONA EXTENDS MILLIONS OF KILOMETRES FROM THE SUN. WHEN ITS MOST ENERGETIC PARTICLES PENETRATE THE EARTH'S MAGNETIC FIELD, THE OXYGEN AND NITROGEN IN THE AIR ARE IONIZED, CREATING BEAUTIFUL AURORAE IN THE POLAR SKIES.

gate rate for billions of years. Over 4.55 billion years, the Sun's energy expenditure has changed little, and its stability is undoubtedly the reason for the appearance and survival of life on Earth.

The Sun is an enormous sphere of gas, 1.4 million kilometres in diameter. Its surface, which appears extremely well-defined, is known as the photosphere. It is the first transparent layer of gas, which transmits the radiation created in the core of the star. In fact, the Sun does not end at this apparently perfect sphere. Observed with a telescope fitted with an appropriate filter, the photosphere looks like a boiling liquid, honeycombed with millions of small regular cells of gas, each as large as France, which emerge and are immediately swallowed up again by the Sun. Above the photosphere there are two other gaseous layers, both transparent and highly rarefied. These are the chromosphere and the corona. These layers of the solar 'atmosphere' are difficult to observe, because they are generally lost in the blinding blaze of light from the photosphere. On Mercury or the Moon, where there is no atmosphere to diffuse the light, the chromosphere and corona must appear whenever the disk of the Sun disappears behind a distant rock. On Earth, where the blue sky is too bright, being able to admire the corona is an exploit in itself, because you need to take advantage of an extremely rare event – a solar eclipse.

The mass of the Sun – two billion billion billion tonnes (or 2×10^{27} t) – is about one thousand times greater than that of all the planets in the Solar System combined. The average density of the Sun

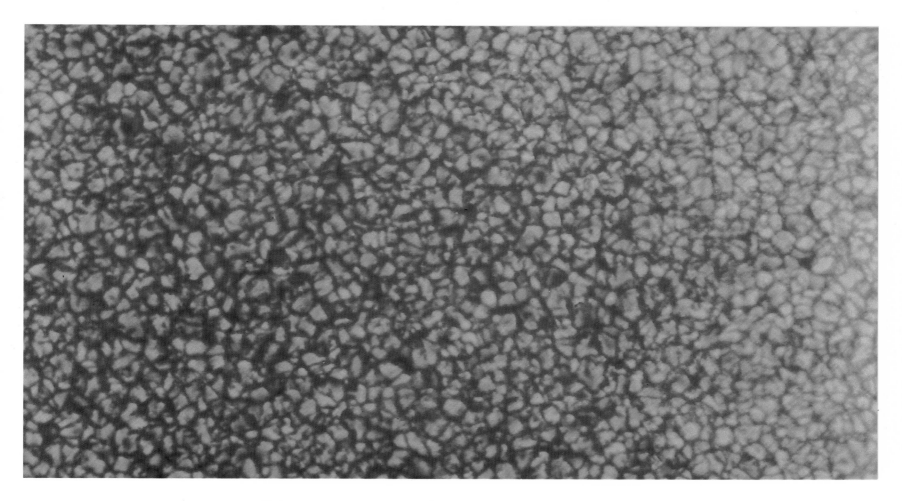

(1.4) is hardly greater than that of water. Its composition is that of most stars: about 70 per cent hydrogen, 28 per cent helium, and 2 per cent heavy atoms, such as carbon, nitrogen, oxygen, neon, and iron.

The physical conditions to which the Sun's material is subject, change radically as a function of depth. At the star's core, the temperature is about 15 million degrees Celsius, and the density is no greater than the extraordinarily low value of 150. To put it another way, one litre of material from the centre of the Sun, brought to Earth, would 'only' weigh 150 kg. By comparison, the same quantity of gold on our planet weighs 20 kg. The gas, tossed around and slowly transported towards the surface, gradually cools, reaching 5500 °C at the photosphere, 700 000 km from the centre. Curiously, despite the apparently thick, opaque, and solid surface of the visible surface of our star, it is actually an extremely rarefied medium that is one thousandth (or even less) of the density of Earth's atmosphere!

For more than four billion years, the flow of radiation from our star has remained, overall, the same. However, despite its appearance of being a perfectly controlled nuclear reactor, the mechanism of the Sun is slightly variable. Every eleven years, the appearance of the Sun changes dramatically. While the visible surface – the photosphere – becomes covered in dark spots, far above, the corona, which is normally elongated over the equator, becomes almost perfectly spherical for several months. These large-amplitude changes are caused by the intense activity of our star's magnetic field. The latter may be represented as a dipole – a simple bar magnet – aligned with the Sun's axis of rotation. But the body of the Sun, which is basically fluid, rotates faster in the equatorial regions than at the poles. In effect, the magnetic lines of force are progressively deformed by this differential rotation. At the time of maximum solar activity, such as that we experienced for several months at the beginning of the year 2000, these lines of force literally burst through the surface of the photosphere. At the points where a tube of magnetic flux enters and leaves the surface, two dark spots of opposite polarity appear. Sunspots are shallow depressions in the otherwise uniformly glowing surface of the photosphere. They are the sites of powerful magnetic activity of the order of several million gauss, which is thousands of times the strength of the Earth's magnetic field. In the centre of spots, which gradually expand until they may measure as much as several tens of thousands of kilometres across, the solar gas is cooler by more than 1000 °C relative to the surrounding photosphere. This difference in temperature explains why, by contrast, the spots appear much darker. After several weeks, the magnetic flux at the centre of the spots declines, and they are gradually engulfed by the cells of searingly hot gas rising from the deeper layers of the Sun. Every cycle, the solar dipole changes sense: the positive and negative polarities switch from one pole to the other. After two complete cycles (i.e., every 22 years), the polarity has reverted to its original state.

To gain a better understanding of the subtleties of the mechanisms operating in the Sun – how the energy emitted by the core reaches the surface; how the rarefied medium of the corona is raised to a temperature of several million degrees – in December 1995, scientists at the European Space Agency launched the Soho satellite, which was equipped with a whole battery of telescopes, capable of observing the Sun at different depths of the photosphere, chromosphere and corona. Never before had the Sun been examined and followed under such favourable conditions. ESA's solar observatory, sited 1.5 million kilometres from Earth, is permanently facing the Sun,

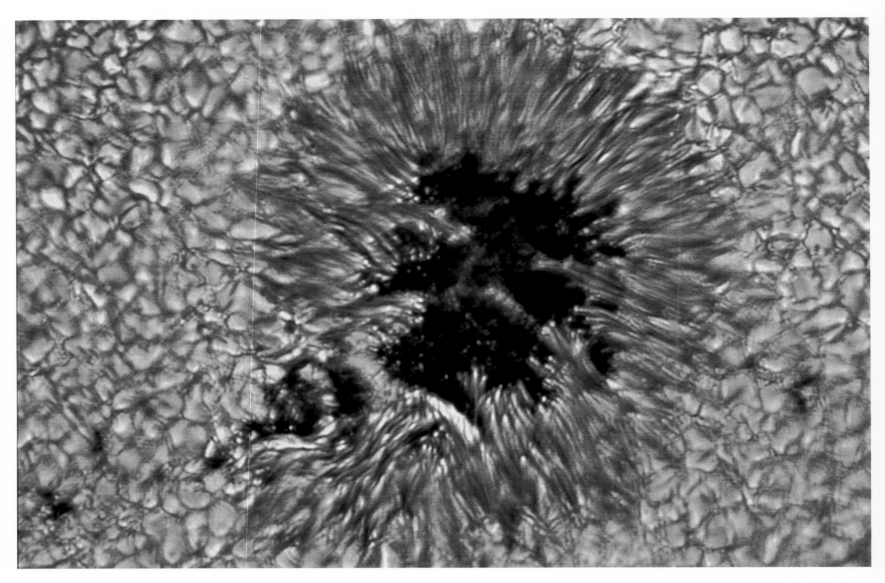

■ Astronomers at Sacramento Peak Observatory in New Mexico obtained this spectacular image of a sunspot in 1998, at about the time the Sun's activity began to increase once more. This spot, which is larger than the Earth, is actually a cooler area of the solar surface.

observing it 24 hours a day. Its telescopes have, among other observations, been able to obtain films – sequences of images – showing, in unprecedented detail, the evolution of material at the surface of the photosphere, as well as the broad motion of plasma deep within the corona. Early in the year 2001, Soho was still functioning. It was thus able to study our star at the time of solar maximum.

The Sun will not continue to provide light and heat for the Solar System *ad infinitum*. Imperceptibly, the properties of the Sun's gases are changing. After 4.55 billion years, the Sun has become impoverished in hydrogen and enriched in helium. Currently, thermonuclear fusion continues in the core at the same rate, a rate that our star can maintain for another 4 or 5 billion years. At the end of this period, the Sun with have almost completely converted the hydrogen in its centre into helium, and its production of nuclear energy will slow down. In fact, the fantastically high gas pressure created by thermo-nuclear fusion will decline, and the helium core, which will then be subject to the force of gravity alone, will slowly begin to collapse. This contraction will, in turn, produce high enough temperatures and pressures for new nuclear reactions to begin. At about 100 million degrees Celsius, helium will begin to be converted into carbon, followed, at about 200 million degrees, by the conversion of helium and carbon into oxygen. The contraction of the core will be accompanied by an expansion of the various overlying layers. The Sun will be transformed into a gigantic body, exceeding 100 million kilometres in diameter, and visible throughout the Galaxy. It will become a red giant, a thousand times as luminous as the Sun today. From then on an irreversible process will begin, which will mark the end of the Sun's adventure. The giant star will expand until it engulfs the closest of the planets, Mercury and Venus. On Earth – but will there be any life on Earth in 5 billion years? – the storm of radiation from the Sun will vaporize the atmosphere and the oceans in an instant, and char mountains and continents. Every morning, for nearly a billion years, a blood-red sphere surrounded by flames will rise above a desert of ashes. Then, succumbing to more and more powerful internal constraints, the Sun will become unstable, and begin to pulsate slowly, like a tired heart. At each of its deep breaths, its prominences will lick the surface of our dead planet. Finally, and suddenly, the Sun will explode annihilating its cortege of planets as it does so, and casting to the Galaxy's four winds the new material that it has created over 10 billion years, as seeds for a future generation of stars.

Mercury:
baked by the heat of the Sun

■ THIS AIRLESS PLANET RESEMBLES THE MOON. LIKE IT, MERCURY IS POCK-MARKED BY IMPACT CRATERS SOME THREE TO FOUR BILLION YEARS OLD. LYING SO CLOSE TO THE SUN, MERCURY IS SUBJECT TO AN INTENSE BOMBARDMENT BY ENERGETIC PARTICLES, WHICH SLOWLY ERODE THE LANDSCAPE. THE PHYSICAL CONDITIONS REIGNING AT ITS SURFACE ARE UNDOUBTEDLY THE MOST EXTREME ANYWHERE IN THE SOLAR SYSTEM: BETWEEN DAY AND NIGHT, THE TEMPERATURE VARIES BY SOME 600 °C.

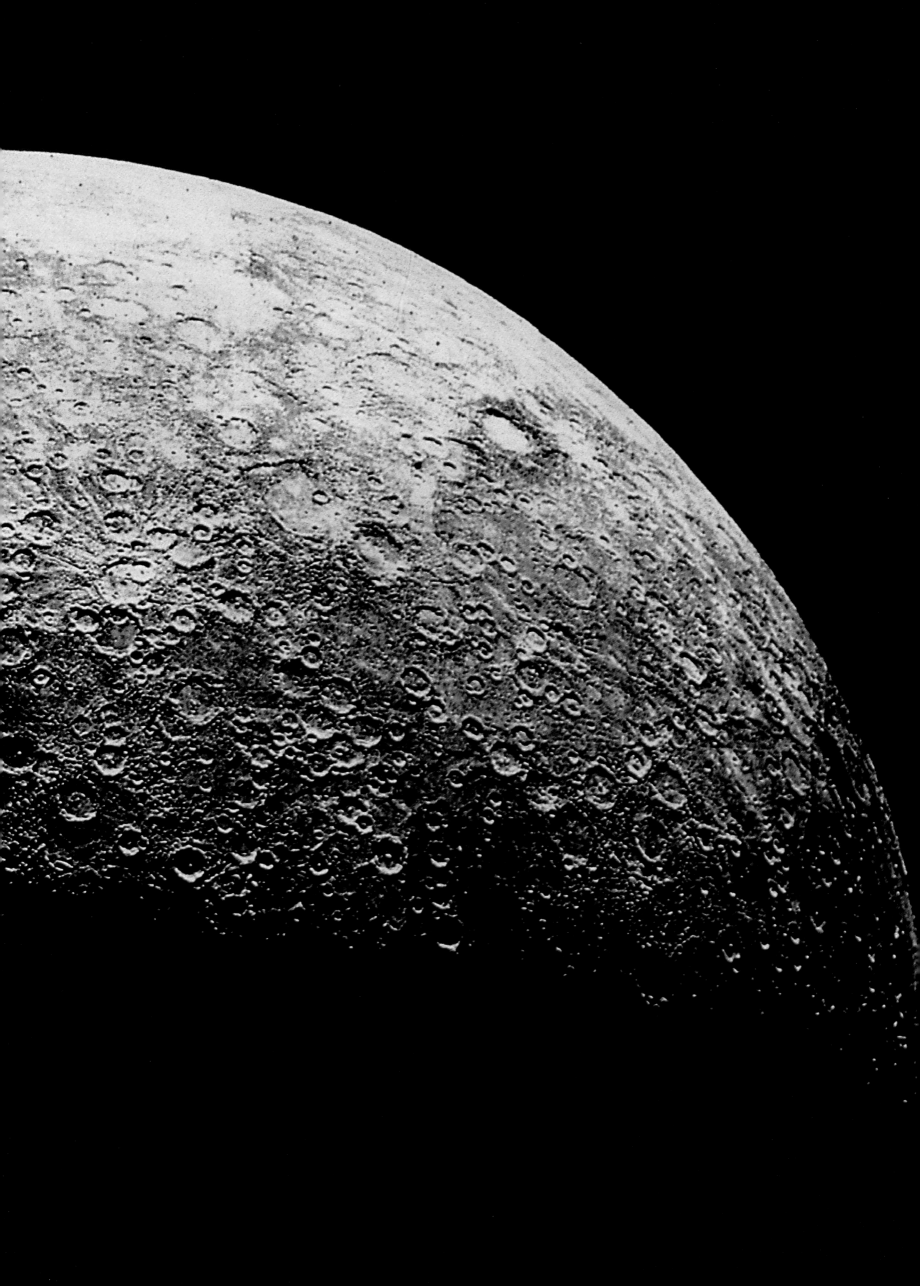

■ THE PLANET MERCURY IS THE CLOSEST BODY TO THE SUN. FOLLOWING A VERY ECCENTRIC ORBIT, IT REGULARLY APPROACHES TO JUST 46 MILLION KILOMETRES FROM THE STAR.

The sky, the colour of ebony, does not show a single star; it seems that the blinding, intolerable light of the Sun is the only thing in existence. At the zenith, the star, far too close, emits an unbearable, inhuman heat. Across the undulating landscape, blasted by the Sun, there is not the smallest pebble, no patch of shade. Everywhere, the same powered ash is dusted over the copper-coloured hills. Here and there, minute craters pock-mark the plain. Beneath the light of Earth, the great expanse of Mare Caloris would be pale and grey, but here the Sun makes it appear blinding like a great expanse of chalk or snow at midday.

The Sun rose on Mercury a month ago. It will not set for another two months. It is impossible to look at the sky: the daytime star, four times as large as on Earth, bathes Mercury in ten times as much energy. Because there is no atmosphere, all the most dangerous radiation – gamma rays, X-rays, and ultraviolet light – drown this small planet in a flood of deadly energy. The place is as uninhabitable as the middle of a nuclear reactor. The relief is muted; rocks crumble into dust. During the day, the temperature in the Sun is close to 430 °C. Placed on the ashen surface of Mercury, a lump of lead or tin would melt in just a few minutes. The following night the thermometer will drop with breathtaking speed, as the planet radiates to space the heat that it has stored during the day. After a month of total darkness the ground will have frozen to −170 °C. Between day and night, the temperature of the ground varies by some 600 °C. No other planet in our Solar System suffers from such extreme temperature variations at its surface.

Mercury is the closest planet to the Sun. Moving at about 50 km/s, it traces out its complete, very eccentric orbit around the Sun in just 88 days: that's the length of its year. One rotation of Mercury on its axis takes 59 days, that is, exactly two-thirds of its period of orbital revolution. Because of this very specific spin-orbit coupling, the true period of time over which the surface of Mercury is exposed to the Sun lasts about three months. At aphelion, the point of the orbit that is farthest from the Sun, the Sun–Mercury distance reaches 70 million kilometres. At perihelion, the closest point to the Sun, the distance is just 46 million kilometres!

Mercury is one of the smallest planets in our system, with a diameter of just 4880 km – about one third of that of the Earth.

■ SOME POLAR AREAS ON MERCURY ARE PERPETUALLY IN DARKNESS, WITHIN THE SHADOWS OF CRATERS. THESE VERY SPECIFIC ZONES, WHICH ARE NEVER HEATED BY THE RAYS OF THE SUN, ARE EXTREMELY COLD AT ABOUT −170°C. IN 1992, ASTRONOMERS BELIEVED THAT THEY DETECTED LARGE QUANTITIES OF ICE IN THESE PARTICULAR AREAS.

■ THIS PHOTOGRAPH SHOWS STRINDBERG, ONE OF THE LARGEST CRATERS ON MERCURY, WITH A DIAMETER OF 165 KM. ABOVE AND TO THE RIGHT IS ANOTHER CRATER, AHMAD BABA, WHICH MEASURES 115 KM. THE LARGE IMPACT FEATURES EXHIBIT AN INTERNAL RINGED STRUCTURE, EVIDENCE OF THE REBOUND THAT OCCURRED AFTER THE INITIAL SHOCK AS THE PLANETOIDS HIT MERCURY.

Although larger than Pluto, Triton, the Moon, Io, and Europa, Mercury is smaller than Ganymede and Titan. Its diameter is almost the same as that of Callisto, one of Jupiter's major satellites, but there is a great difference in mass: it is three times as great as that of Callisto. This is because Mercury, with the Earth, holds the record for density: 5.45. For comparison, the density of the Earth is 5.52; the Moon, 3.34; Callisto, 1.80; and that of certain small satellites of Saturn is around 1.0 – in other words, the density of water!

Mercury undoubtedly owes the high density of the materials of which it consists to its proximity to the Sun. At the time of the formation of the Solar System, the light elements, such as hydrogen, helium, and oxygen, were blown out to more distant regions, while the heavy atoms, such as the metals and silicates, remained close to the newly born Sun. As a result, Mercury is almost a metallic sphere; its iron-rich core, which is the largest of any of the bodies in the Solar System relative to its mass, has a diameter of more than 3500 km, as against the planet's total diameter of 4880 km. When Mercury was formed, the heavy materials were molten and flowed towards the centre to create the core – which probably amounts to about 80 per cent of the mass of the planet – and a lighter layer of silicates was formed at the surface.

The surface of Mercury is one of the most ancient in the Solar System. Saturated with craters, to the extent that the impacts overlap, in places it shows smoother plains, which bear witness to the spread of lava flows that date back almost four billion years, as in the case of the Caloris Basin, one of the youngest regions on Mercury. This basin, 1200 km across, was created by the exceptionally violent impact of an asteroid, several tens of kilometres in diameter, which happened 3.9 billion years ago and affected the whole planet. The shock created concentric mountainous rings, 2000 m high, and the seismic waves travelled throughout the planet causing it to ring like a bell. At the point at which they converged – i.e., at the antipodes of the impact – the surface was devastated. It is now the site of a chaotic terrain that is covered with ridges and fractures, and where, in certain places, ancient impact craters are unrecognizable, having been disrupted and wiped out by the ferocity of an 'earthquake' that was perhaps a thousand times as violent as those that affect our own planet today.

Mercury's only visit has been by the American Mariner 10 spaceprobe, which flew past it three times during 1974–5, before failing and becoming trapped in a permanent solar orbit. No robot from Earth has landed on Mercury to take photographs of the landscape. The exceptionally severe thermal constraints to which it would be subject have, so far, discouraged the major space agencies.

But scientists most definitely want to return to Mercury. First, in 2004, NASA is going to launch the Messenger probe, and this should go into orbit in 2009, beginning the long task of cartography. The resolution of the images obtained by this American probe will depend on its altitude at any one time: about ten metres at perigee (200 km altitude), but only 1 km at apogee (15000 km). The European Beppi-Colombo probe is far more ambitious. In 2011 the ESA probe will enter a far less eccentric orbit than Messenger. There, at altitudes of only 300–600 km, the probe will photograph details at a resolution of 10–20 m over the whole of the planet. Beppi-Colombo also carries a small landing module that will be released to study the searingly-hot surface directly. This module, equipped with a mini-rover and a 'mole', will provide a better understanding of the mineralogical and chemical composition of the surface and will, of course, obtain a full photographic panorama of the landscape.

It is the end of the afternoon on Mercury, after a long day in the sunlight, which has lasted nearly three Earth months. The shadows cast by the red-hot rocks become fantastically

elongated, and the temperature begins – relatively – to fall. The twilight is a spectacle of unbelievable beauty. After taking 48 hours to set, the blinding disk of the Sun has finally been hidden behind the line of distant mountains. But the star's presence persists for hundreds of hours with a resplendent view of the corona, stretching some 4–5° across the sky, with its complex, changing rays and streamers. This glorious sight may, sometimes, be accompanied, at the very moment the Sun sets, by eruptive prominences, delicately pink in colour, which slowly rise into the dark sky by perhaps as much as a degree, before sinking below the horizon, either carried by the Sun's apparent motion, or because they have sunk back towards the surface. Much higher in the sky, a vast, faint area of light fades away into space. This is the zodiacal light, a transitory cloud of interplanetary dust particles, which scatters the radiation from the Sun and occupies the main plane of the Solar System. Neither by night nor by day is there the slightest perceptible trace of any atmosphere on Mercury. Except for the ghostly zodiacal light and the Milky Way, the sky is perfectly dark, even immediately above the horizon. Yet infinitesimal traces of gases have been detected at Mercury's surface. These gases, primarily consisting of helium, undoubtedly derive from the solar wind, trapped for a short time by Mercury's gravitational field.

Finally night descends. Having sweltered in the sunlight for three months, Mercury's surface cools down with incredible rapidity: in just a few tens of minutes the thermometer stabilizes at around −170 °C. Darkness will last three months. But, once night has fallen, Mercury offers the finest vantage point over the whole Solar System that can be imagined. Visible to the naked eye, and rising over the eastern horizon one after the other, are Venus, the Earth, Mars, Jupiter, and Saturn. These five brilliant, beautifully coloured planets shine with a steady light in the sky. Jupiter and Saturn, yellowish and subdued, resemble the view they present in the Earth's sky. Mars, distinctly reddish, is not much fainter than Saturn. The eye is drawn irresistibly to the Earth, which looks like a beautiful turquoise, sparkling in the sunlight. Alongside it, minute and dark, the Moon is almost invisible. But the Mercurian night is dominated by the unreal, disquieting light of Venus, fully illuminated by the Sun. Its yellow brilliance is so intense that it casts weak shadows on Mercury's desolate surface.

Venus:
a vision of Hell

■ In appearance, the beautiful planet Venus is a twin sister to the Earth, with whom it shares the same size, the same mass, and the same chemical composition. Yet, closer to the Sun than our blue planet, Venus has suffered a completely different fate. Eternally shrouded in thick clouds, it hides a surface that is a desert, crushed beneath a horrendously dense, burning atmosphere, where humans will probably never set foot.

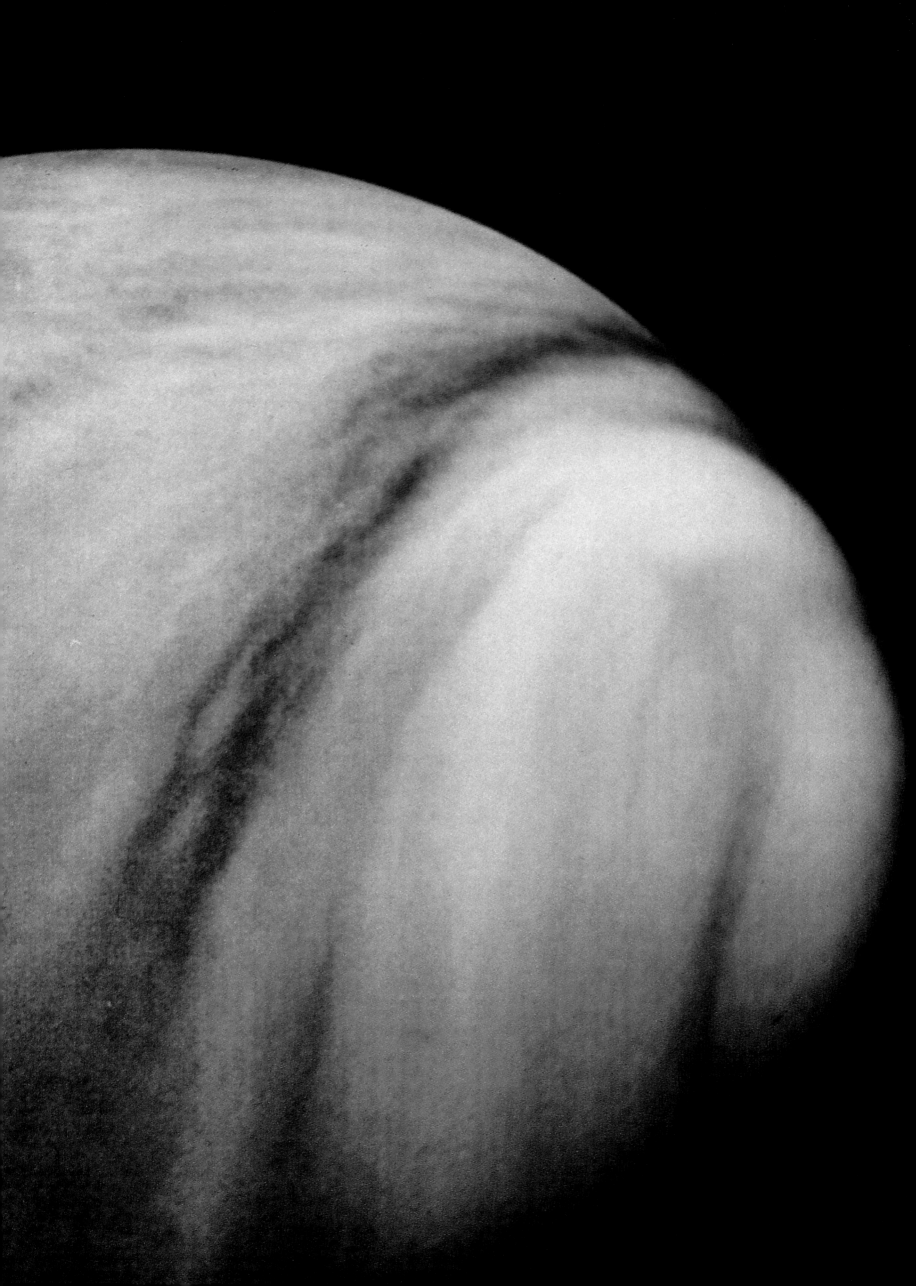

■ Several million years ago, a lava flow crossed the Denitsa region on Venus. The flood of extremely fluid lava eroded its bed to a width of 1–4 km over a distance of more than 120 km.

The fog and sky seem to merge. The landscape here is indistinct, coming and going at the whim of the slowly swirling clouds. The ground looks as if it has been paved, covered with large and small flagstones, which soon become lost in the mist. The atmosphere is calm, static, without the slightest breath of air. In the distance, the dull rumble of thunder is heard from an invisible storm. The horizon is lost in the veil of fog, and the rocky soil is an orange colour, sprinkled with coarse brown sand. The Sun is hidden, but the light from the sky is diffuse, uniform, yellowish, and too weak to cast any shadows. Dawn broke about two Earth months ago on the planet Venus, and night will not arrive for another two months. The light will then fade slowly, and it will take dozens of hours for twilight to fall. The dim silhouette of the surrounding low mountains will be seen for some time against the sky as it gradually darkens to grey. And then it will be night. Utter darkness, with absolute silence, occasionally broken by a distant rumble of thunder. Four months of the blackest night, and the clouds never break to allow even the slightest glimmer of starlight to filter through.

Four months without sunlight, and yet the temperature will not drop. On the sand-strewn surface, on rocky flagstones broken by the shock of its landing, sits a giant metallic, sooty beetle. Its glittering carapace is slowly corroding. This giant insect from space has been lying here since Earth year 1982. Before it, twenty-five spaceprobes had tried to penetrate the secrets of Venus. Most did not reach the ground in a working state. But Venera 13 and Venera 14 did succeed in doing so. For about an hour they sent radio messages back to Earth, and had sufficient time, before expiring, to warn us that Venus is Hell.

Venus is the second planet from the Sun. It revolves in the most regular of all the planetary orbits, following an almost perfect circle around the Sun. At its minimum distance, at perihelion, it approaches the star to a distance of 107 million kilometres; and at aphelion it only moves out to 109 million. A Venerian year lasts 225 days. Venus, like Mercury, does not have even the tiniest satellite. Like Mercury and the Earth, this rocky planet has a very high density of 5.25, in contrast to 5.45 and 5.52 for the other two planets. At first glance, Venus is Earth's twin sister. It has the same density; almost the same diameter, 12 100 km, as against 12 756 km for Earth; it too has an atmosphere; and, finally, it is on a neighbouring orbit. In theory, these two bodies should therefore have very similar

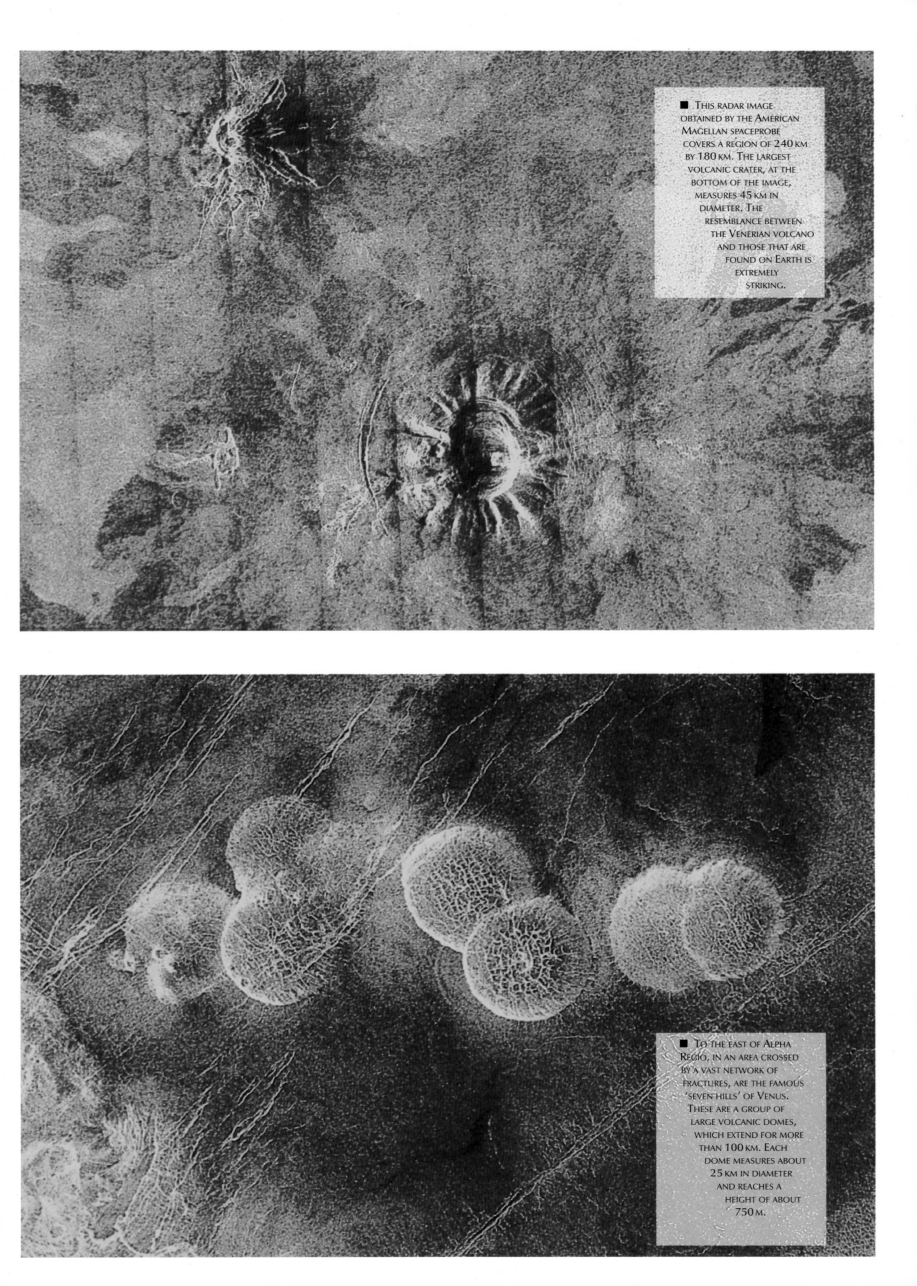

■ THIS RADAR IMAGE
OBTAINED BY THE AMERICAN
MAGELLAN SPACEPROBE
COVERS A REGION OF 240 KM
BY 180 KM. THE LARGEST
VOLCANIC CRATER, AT THE
BOTTOM OF THE IMAGE,
MEASURES 45 KM IN
DIAMETER. THE
RESEMBLANCE BETWEEN
THE VENERIAN VOLCANO
AND THOSE THAT ARE
FOUND ON EARTH IS
EXTREMELY
STRIKING.

■ TO THE EAST OF ALPHA
REGIO, IN AN AREA CROSSED
BY A VAST NETWORK OF
FRACTURES, ARE THE FAMOUS
'SEVEN HILLS' OF VENUS.
THESE ARE A GROUP OF
LARGE VOLCANIC DOMES,
WHICH EXTEND FOR MORE
THAN 100 KM. EACH
DOME MEASURES ABOUT
25 KM IN DIAMETER
AND REACHES A
HEIGHT OF ABOUT
750 M.

surface conditions. In fact, the Earth's average temperature, if it were at the same distance from the Sun as Venus, would increase by 'only' 20 °C. So, then, shouldn't Venus be a planet of eternal summer, where tropical lagoons fringed by tree-ferns shelter immense, placid reptiles?

The reality is more like Hell: who would dare to visit this nightmare world, which is plunged into perpetual gloom, where the sky never clears enough to let the sunlight through? It has been a suffocating furnace for four billion years, where the temperature, day and night, never drops below 460 °C. In Phoebe Regio, a dreary expanse of rocks enveloped in the yellowish fog that covers the whole planet, the atmosphere is so dense that it seems palpable. At the surface of the planet, which is hotter than that of Mercury, lead and tin would melt spontaneously. And if the air is remarkably still, with never the slightest breath of wind, it is because the Venerian atmosphere is incredibly heavy. The slightest movement here takes place in slow-motion. During Venerian 'storms', which arise particularly around the time that the invisible Sun sets, the most violent gusts reach no more than 1–4 km/h. These silent hurricanes would be capable of lifting you up as easily as any ground swell.

The lower atmosphere of Venus resembles a deep, burning ocean. At the surface of the planet the atmospheric pressure is the same as that at a depth of 1000 m in the Earth's oceans. This atmosphere was so new and unexpected, that the Soviets, who even today remain the only people who have fully succeeded in getting a spaceprobe to function on Venus, lost more than a dozen of them before realizing that the probes needed to be designed as deep-diving submarines, or bathyscaphes, capable of surviving in an ocean. The Venerian atmosphere is so dense that the last wave of Soviet probes did not use parachutes over the last fifty kilometres before reaching the surface! This atmosphere consists of 96 per cent carbon dioxide, less than 4 per cent of nitrogen, and traces of other gases. It is very similar to that of Mars, although 10 000 times as dense! The air is astonishingly dry: there is no water, nor any water vapour here. The average atmospheric pressure is 90 bar, as against 1 bar on Earth, and 0.01 bar on Mars. It is

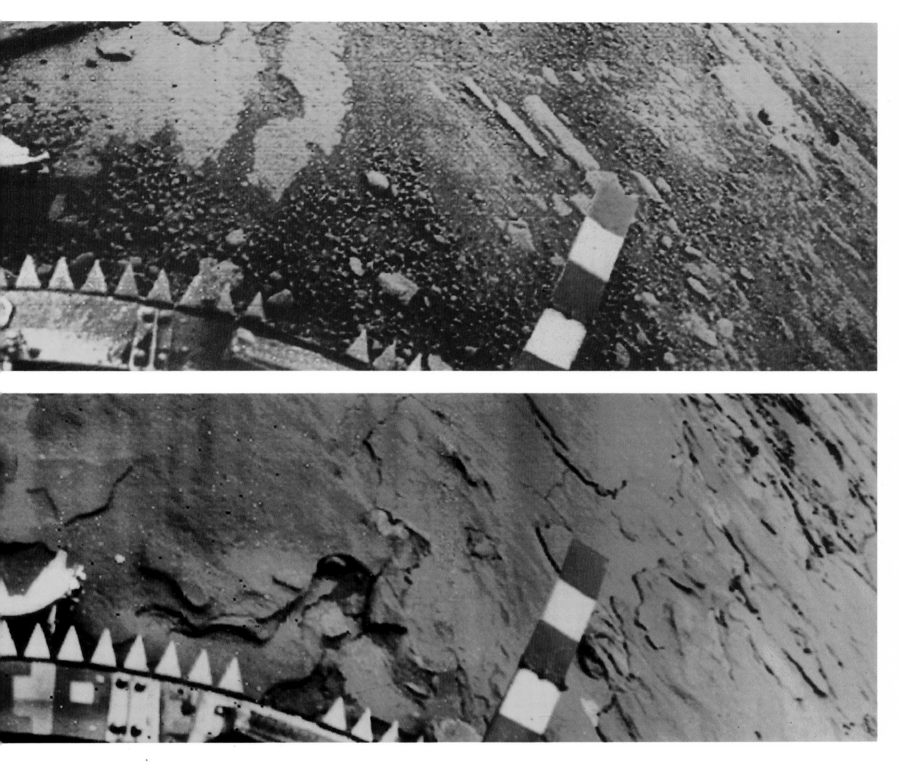

worth pointing out that under this terrible pressure, water, if there were any, would be able to remain liquid up to a temperature of nearly 380 °C! How could a planet so apparently similar to the Earth, suffer such a different fate?

When the Earth and Venus were formed, some 4.55 billion years ago, the two bodies were probably very similar, their surfaces being pelted with asteroids, and releasing the gases that were to become their atmospheres. On Earth, cooler and farther from the Sun, water molecules were able to exist, at first in the form of water vapour, and then condensing into liquid water. Carbon dioxide, expelled in vast quantities by volcanoes, was dissolved in the primordial oceans, and imprisoned in silicate and carbonate rocks such as limestone. Carbon dioxide (CO_2) is a greenhouse gas. The principle behind the greenhouse effect is simple: solar radiation is partially absorbed by the atmosphere and the surface, which heat up. Normally, the planet's surface radiates the heat that it has accumulated during the day into space during the night.

Greenhouse gases are opaque to this infrared radiation, and absorb and store it, so the atmosphere has a certain thermal inertia. The presence of carbon dioxide in trace amounts – just 0.03 per cent – in the Earth's atmosphere has kept the latter at a pleasant temperature, and allowed water to exist in all three of its phases: solid, liquid, and gaseous. The average temperature on Earth is 15 °C. Without CO_2, our planet would be a frozen desert, with a temperature of about −20 °C.

The history of Venus is completely different. From the very beginning, the average temperature at the surface of the planet was higher than that of Earth by some twenty degrees Celsius. The water vapour and carbon dioxide released from the Venerian crust accumulated in the atmosphere, covering it with clouds, and giving rise to a particularly efficient greenhouse effect, which gradually became even more extreme. Were there at one time, shortly after the formation of the planet, when the atmospheric conditions were similar to those on Earth, oceans on Venus? We do not know. If there were, the water must have quickly evaporated, and its molecules broken down. The light hydrogen atoms would

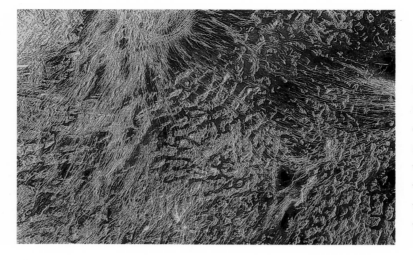

■ THIS PORTION OF OVDA REGIO, THE WHOLE OF WHICH WAS OBSERVED BY THE MAGELLAN PROBE, MEASURES NEARLY 600 KM BY 250 KM. AN INTRICATE NETWORK OF FAULTS AND RIDGES COVERS THIS ANCIENT PLAIN, WHICH APPEARS TO HAVE BEEN INVADED BY LAVA FLOWS. SUCH IMAGES PROVIDE PLANETOLOGISTS INFORMATION ABOUT OVERALL MOVEMENTS OF THE VENERIAN CRUST.

have escaped to the upper atmosphere, and thence to space, leaving the carbon dioxide to cover the planet with an opaque, suffocating blanket. Slowly the temperature at the surface rose. Sunlight should have been able to reach the surface more than three billion years ago. Could Venus perhaps have been a true Paradise during that long-gone age? During the middle of the Archaean era, the Earth developed immense colonies of blue-green algae in its peaceful oceans. What was happening at that time on the neighbouring planet? Because today there is no trace of that far distant past, we shall probably never know. If the whole of the water contained in the atmosphere were to be precipitated onto the surface of Venus, it would form an even layer just one centimetre deep, as against several kilometres on Earth. Over the past four billion years, the temperature of the lower atmosphere of Venus, which is simply a vast ocean of carbon dioxide, has progressively increased, from 50 °C to 100 °C, from 100 °C to 200, 300, and 400 °C. Thermal equilibrium was reached at the current value of 460 °C, and henceforth the thermometer is unlikely to either rise or fall. Hidden beneath an atmosphere that is several tens of kilometres thick, the surface of Venus will never see the Sun again.

Very heavy and dense, the atmosphere of Venus is also very thick. The top of the cloud layer lies at an altitude of about 70 km. It is this cloud cover that astronomers observe, and which spaceprobes, such as Mariner 10 or Pioneer Venus studied in detail in the 1970s. The clouds reflect much of the sunlight. Venus, seen from a distance, such as from the surface of Mercury or from the Earth, is extremely bright. Like the Earth's atmosphere, that of Venus becomes more rarefied with height. On Earth, 90 per cent of the mass of the atmosphere is concentrated in the first 10 kilometres. The air pressure, equal to one bar at sea level, is just 0.5 bar at 5500 m, and 0.3 bar at the summit of Mount Everest, at 8848 m. The air becomes practically

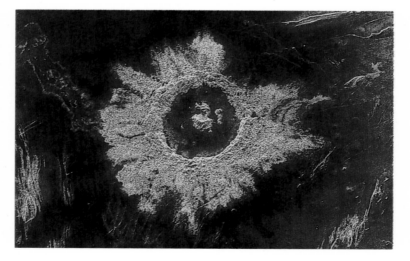

■ CREATED BY THE IMPACT OF AN ASTEROID, SEVERAL HUNDREDS OF MILLIONS OF YEARS AGO, THE CRATER DANILOVA MEASURES 48 KM IN DIAMETER. IT SHOWS THE CHARACTERISTICS TYPICAL OF IMPACTS ON VENUS. BRAKED BY THE EXTREMELY DENSE ATMOSPHERE, THE DEBRIS FELL BACK CLOSE TO THE CRATER AND FORMED LOBES, WHICH APPEAR BRIGHT ON THIS RADAR IMAGE.

unbreathable as soon as one reaches an altitude of 7000 or 8000 m. On Venus, at the summit of the highest mountain range, Maxwell Montes, which peaks 11 000 m above the average level of the surface, the pressure is about 70 bar, and the temperature is around 400 °C. Far above the highest summits, at an altitude of 15 000 m, the pressure is below 40 bar, and the temperature a cool 360 °C.

Higher still, at about 45 km altitude, there is a layer in the Venerian atmosphere where the pressure is exactly the same as it is on Earth: 1 bar. In 1985, the Soviets dropped two balloon probes to this level. Released by the Vega 1 and 2 spaceprobes, which flew past Venus at a distance of 11 000 km, the balloons survived for more than forty hours. This region reserves some nasty surprises for any hypothetical – and foolhardy – travellers: the fog is so dense that visibility is zero; the temperature of the air approaches 100 °C; the winds rage at 300 km/h; and, above all, the precipitation is dreadful – Venerian rain consists of droplets of sulphuric acid!

Not far from the top of the clouds, the temperature finally reaches values that are comparable with those on Earth. At an altitude of 60 km, the air – obviously fatal for human lungs, which would instantly be seared by the clouds of sulphuric acid – is as dense as it is on Earth at an altitude of 5500 m, with a pressure of 0.5 bar and a temperature of 0 °C. Any unlikely inhabitants of Venus, escaping from the planet for the first time, would pass through a totally opaque atmosphere before finally emerging from the clouds, when they would be astounded to discover the sight of a sea of clouds sparkling in the Sun. They would then have reached an altitude of 70 km, and would be soaring in extremely rarefied, freezing air, at a temperature of 40 °C lower.

The atmospheric circulation on Venus is very strange. At ground level, where it is as heavy and dense as molasses, the remarkably still atmosphere is controlled by the rotation of the planet. The latter rotates very slowly, just

once in 243 days, and winds at the surface of Venus, as we have seen, rarely exceed 4 km/h. By contrast, the upper atmosphere, one hundred to one thousand times less dense, is totally independent of the planet, and literally slides over the lower atmospheric layers. It rotates around Venus in just four days, the flow being generally parallel to the equator, with bands of cloud that move towards the north, creating recumbent 'V' and 'Y'-shaped cloud formations visible from outside the planet.

We know the surface of Venus only from the four Soviet Venera probes that succeeded in photographing the desolate landscape where they landed before they were overcome by the crushing, corrosive, and searing atmosphere. The hostility of Venus may be measured by the lifetimes of the probes that landed there. For Veneras 9 and 10, 53 minutes and 65 minutes, respectively. For Veneras 13 and 14, 57 minutes and 127 minutes. In comparison, the Martian probes Viking 1 and 2 (the American equivalents of the Venera probes), were able to function on Mars for three-and-a-half years in the case of the second, and nearly seven years for the first! To finally unveil Venus, in 1989 the Americans launched the Magellan probe. Using a radar that emitted radio waves, the only thing capable of piercing the otherwise impenetrable clouds, Magellan was eventually able to obtain images of the whole of the planet, between 1990 and 1993. The result? A map of the planet as it would

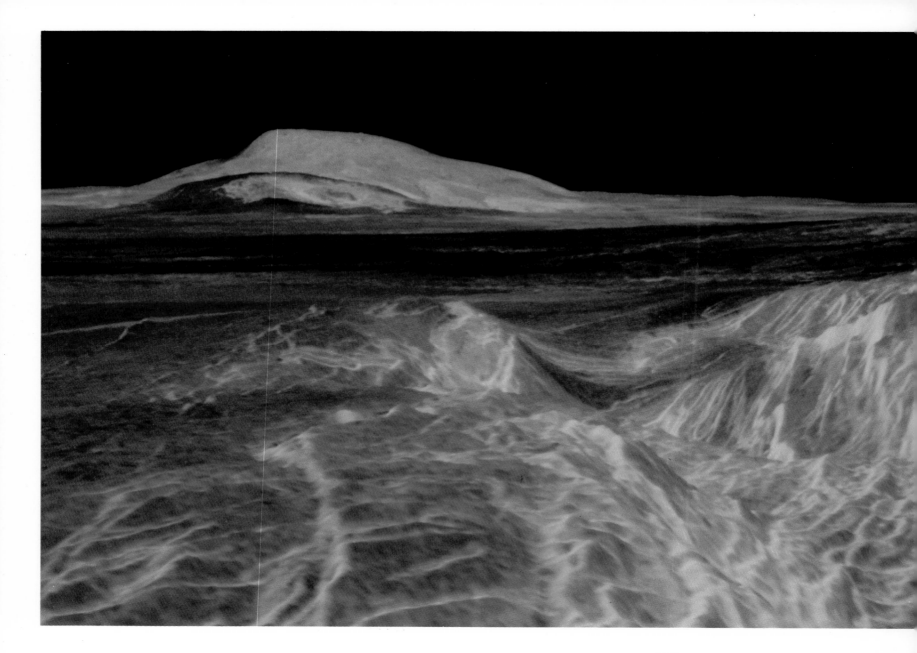

appear from the porthole of a hypothetical spacecraft if, like Mercury, it had no atmosphere at all.

The surface of Venus is an immense monotonous plain, a vast, flat desert of basaltic rocks, with a remarkably even level. About 70 per cent of the surface of the planet is at the same altitude, which has naturally become, in the absence of any sea level, the reference level for the planet. Low hills and plateaux a few hundred metres high are scattered here and there across the desert, which is also criss-crossed by numerous sets of faults, both parallel to, and at right-angles to one another, creating a fine network of gorges and ridges, witness to the planet's intense internal activity. Three great mountain massifs emerge from this desolate plain: Aphrodite Terra, which extends for about 3000 km along the equator, and rises to a height of more than 5000 m; Beta Regio, at latitude 30° north, with a height of 4000 m; and finally, not far from the north pole of Venus, Ishtar Terra, a particularly high, sharply defined plateau, as large as a terrestrial continent. This plateau rises about 5000 metres above the general level of the planet. Some experts see it as an attempt by Venus to develop a system of plate tectonics, like that on Earth. Ishtar Terra includes one of the largest mountains in the Solar System, Maxwell Montes, which is 750 km across and 11 000 m high. Not far from the summit there is a crater 105 km in diameter, known as Cleopatra, from which runs the bed of an ancient lava flow. This crater was formed several hundred million years ago by the impact of an asteroid. The crust of Venus, liquefied, then spread out over the flanks of the mountain. Compared with Mercury, the Moon, or Mars, there are very few craters on Venus. A few hundred traces of impacts have been detected, which represents one crater per million square kilometres. This is a cratering rate one thousand times less than that of Mercury or the Moon! No craters have diameters of less than 2 km – a minimum size that is explained by the thick atmosphere of Venus, which acts as an extremely effective protective screen. On the Moon, and even on Earth, meteorites of just a few tenths of a metre or a few metres across can reach the surface and excavate 'bomb craters' some metres in diameter. On Venus, the smallest meteoroids arriving from space are destroyed as they pass through the atmosphere. Most arrive at the surface only in the form of a meteoritic dust attacked by sulphuric acid. From the observed craters, probably only asteroids that are larger than 50 m across are capable of entering the Venerian furnace. The ejecta created by the impacts are themselves very unusual. They form thick lobes, because the material ejected by the impacts was braked by the planet's dense atmosphere. But the atmospheric screen does not explain the rarity of impact craters with diameters greater than 2 km. In fact, the surface of

EXCEEDS 3000 M. THE RIDGES IN THE FOREGROUND ARE
PERHAPS ANCIENT LAVA FLOWS THAT SOLIDIFIED LONG
AGO. ON THE LEFT, THERE IS A SHIELD VOLCANO, SIF
MONS, WHICH RISES TO A HEIGHT OF 2000 M.

Venus is geologically very young, and is undoubtedly still active today. From the north to the south, the planet is covered inextricably in a network of ridges and faults, witness to complex cycles of compression and extension of the crust.

At the planet's equator, hundreds of strangely shaped volcanic edifices are found. In Alpha Regio, seven large, cracked domes, 25 km in diameter have tiny vents at their summits, which are no more than 750 m high. Farther away, to the west of Ishtar Terra, there is Sacajawea Patera, an immense cirque, surrounded by a ring of mountains. This depression, more than 200 km in diameter and 2000 m deep, lies above an ancient, subterranean reservoir of lava. Finally, all over the planet, there are the sinuous beds of ancient lava flows, between 500 and 1500 m wide, weaving round mountains and filling valleys. These former rivers of fire meander over distances of tens or even hundreds of kilometres. And it was on Venus that the longest flow of lava in the whole Solar System was found: in Niobe Planitia, it stretches for more than 7000 km!

With few craters, and reworked by incessant volcanic activity, the surface of Venus is undoubtedly less than 500 million years old. What's more, planetologists are convinced that this volcanic activity still continues today, even though Magellan has not directly imaged any volcano in eruption or any current lava flow.

Venus, despite its astonishing apparent resemblance to the Earth, has been found to be a completely different world. A small difference in distance from the Sun has been sufficient to transform the planet into an unlikely hell. These desolate landscapes, with worn-down rocks, bathed in a suffocating, crushing atmosphere, have no equivalent anywhere else in the Solar System. If, one day, some incredibly equipped astronaut were to tread, with strangely slow movements, on this other-worldly desert, he would doubtless ponder, not without a feeling of uncertainty, on the fact that our own planet might, after all, have suffered the same fate.

The Earth:
the story of a living planet

THE EARTH IS UNDOUBTEDLY THE JEWEL OF THE SOLAR
SYSTEM. ADMIRING IT FROM SPACE, THE NAME SEEMS TO
HAVE BEEN POORLY CHOSEN FOR THIS SOFT, BLUE BALL
FLOATING IN INFINITE SPACE. IN FACT, THREE-QUARTERS OF
THE SURFACE OF OUR PLANET IS COVERED BY THE OCEANS.
THE PRESENCE OF LIQUID WATER ON A PLANET IS
POSSIBLY AN EXCEPTIONALLY RARE EVENT IN THE
UNIVERSE. AND WHAT ABOUT LIFE?

■ In the far north of the Earth, the glaciers on the Kamchatka peninsula fall directly into the sea. Ice, water, and clouds show the precious molecule of H_2O in all its three states.

The wind had increased yet again. The powdery snow was flying violently across the royal-blue sky, and the whole mountain seemed to smoke. The icy wind which had been blowing continuously for several weeks had scalloped the summit of Machhapuchhare, and the cornices of ice that ran along its summit ridge looked like dazzling waves, frozen in time. It is winter on Nepal's sacred mountain. At an altitude of 7000 m, a thermometer would indicate a temperature of −50 °C. But no one has yet set foot on the summit of the most beautiful mountain on Earth. And who, or what, could live here? A few square metres of blue ice covered with hardened snow hide the granite. A sublime vantage point overlooking the high chain of the Himalayas to the north, and the hot, humid plains of India to the south, this tiny platform of ice, between heaven and earth, is perhaps more inhospitable than the planet Mars.

The same planet, but much farther west. The Sun, directly overhead, washes out all shadows and sears the landscape. Here, the sky is white, the horizon is hazy, indistinct, and lost in mirages. There is not a breath of wind and nothing but burning sand and the frozen waves of dunes as far as the eye

can see. This is the deepest desert. There is not a single track visible on the sand, and the ambient temperature at ground level has reached nearly 60 °C. In the distance, the dunes and rocks seem to dance in the suffocating air. Nothing moves, except perhaps the Sun, which slowly follows its path towards the west. Yet again, what visitor to this lost corner of the central Sahara, at the bottom of the Ouan Rechla basin, would believe that the Earth is an inhabited planet? Yet over there, hidden by a ridge of dark rocks, which protect it from erosion by the winds, there is a large fresco, carved into the rocks 8000 years ago, waiting to be discovered. It portrays a fauna of large animals, hunting scenes, and obscure rituals. The mature craftsmanship, mingling figurative representation with abstract themes, and the delicacy of the lines bear witness to a fully evolved species, conscious of its own existence. Farther on, in a rock shelter, paintings that are just as old are slowly disappearing, being buried beneath the sand. Has life deserted this planet?

Far from the Sahara and the Himalayas, and even farther to the west, a giant cumulonimbus storm cloud begins to swell, rising slowly in the heavy atmosphere, charged with electricity.

■ THE INACCESSIBLE SUMMIT OF MACHHAPUCHHARE, AT AN ALTITUDE OF 7000 M. NEPAL'S SACRED MOUNTAIN, DUSTED IN WHITE BY THE MONSOON, DOMINATES THE GORGES OF THE MODI KHOLA RIVER, WHICH FEEDS THE KALI GANDAKHI, WHICH IS ITSELF A TRIBUTARY OF THE GANGES. EVER SINCE THE DAWN OF TIME, MOUNTAINS AND RIVERS HAVE SHAPED THE SURFACE OF THE EARTH, AND ARE A MONUMENT TO THE PERPETUAL CYCLE OF WATER ON OUR PLANET.

■ FROM A DISTANCE OF
150 000 KM, WHEN *EN ROUTE*
FOR THE MOON, THE
ASTRONAUTS OF THE APOLLO
11 MISSION TOOK THIS
SPECTACULAR PHOTOGRAPH OF
THEIR HOME PLANET. BEHIND
THE SWIRLS OF CLOUD THAT
COVER PART OF THE EARTH,
VAST AREAS OF AFRICA,
EUROPE, AND ASIA ARE
VISIBLE.

The wind from the south has dropped and, for the third time today, rain has started to fall. The temperature is 29 °C and the humidity is 98 per cent. A weak rainbow appears in the east, through the persistent drizzle. With its veils of mist and rays of sunlight, here the day will soon come to an end. Nothing moves. At least, that's how it seems, but beneath the foliage of this virgin rain-forest there are millions of insects and arthropods, scorpions, tarantulas and giant beetles; toucans, macaws, and ibis; caymans, boa constrictors, anacondas, iguanas, and tortoises; piranhas; jaguars, sloths, tapirs, and armadillos – all are hiding, fleeing, and hunting in the Amazonian mangroves as they are overtaken by the night.

We are on the third planet from the Sun. From its general appearance, nothing distinguishes this body from the others in the Solar System. This planet is not the largest, nor the smallest. Although it is not surrounded by a system of rings, like the giants, it does have a large satellite. It is also perfectly ordinary from the physical and chemical point of view, sharing many of its characteristics with the other bodies in the Solar System. Its diameter is 12 756 km, practically the same as Venus. Its density is 5.52, almost the same as those of Venus and Mercury. It chemical composition and internal structure are identical, in all major respects, with those of the Moon, Mercury, Venus and Mars. Its dynamical properties are such that, like most of the rest of the planets, the Earth revolves around the Sun in an orbit that is practically circular, lying close to the mean plane of the Solar System. Its rotation on its axis takes twenty-four hours, almost identical to that for Mars. Like Mars also, its rotational axis is tilted by about 24°.

Yet the Earth is different. Its atmosphere, less dense than that of Venus, is much thicker than that of Mars and roughly resembles that of mysterious Titan. But the composition is unique in the Solar System: 78 per cent nitrogen, 21 per cent oxygen, and less than 1 per cent carbon dioxide, argon, methane, ozone, etc. And, above all, at its average pressure of 1013 millibars, the terrestrial atmosphere shelters the most promising molecule in the universe – H_2O. Two atoms of hydrogen bonded to one of oxygen – water. This mole-cule is not rare in the Solar System, it is found on Mars, Europa, Enceladus, and elsewhere; on the surface of comets, in the rings of Saturn ... more or less everywhere in fact, but always in the form of ice, hard, solid, immutable, and sterile. Chance has meant that the third planet occupies a unique place in the Solar System: the Earth's orbit lies precisely within an extra-ordinarily narrow torus, where liquid water can exist. At its minimum distance, the Earth is 147 million kilometres from the Sun; at its maximum distance it is 152 million kilometres away. Closer by some 15 million kilometres, and the Earth would have drifted towards becoming the hell that is Venus. Some 15 million kilometres farther out, and the Earth would long ago have become a world of ice. This ideal distance from the Sun, linked with an ideal mass, and an atmosphere that has the ideal density, has meant that the Earth is the jewel of the Solar System. The blue planet, the ocean planet, the living planet: the Earth never ceases to change, to evolve, and to be reinvigorated thanks to the influence of this precious, powerful, fluid.

Although the Earth is the same age as the Solar System, 4.55 billion years, it is also, in the geological sense of the term, the youngest – along with Io – of all the bodies that orbit our star. Here, everything started just as it did on the other planets. In the nascent Solar System, the recently formed Earth, searingly hot, surrounded by a dense, hot, and opaque atmosphere, underwent an incessant bombardment by the meteoroids that were criss-crossing space in all directions. A core of iron and nickel formed at the centre of the Earth, around it appeared the silicate mantle, and, finally, the crust, as thin as the shell on an egg, and only a few tens of kilometres thick. Millions of volcanoes spat clouds of carbon dioxide, nitrogen, and water vapour into the cloud-covered sky. The primordial ocean formed in various low-lying basins and at the bottom of craters, and was only prevented from boiling and evaporating by the enormous pressure. Magma emerged from the depths through immense scars in the terrestrial crust, which was constantly pounded by the fall of aster-oidal fragments.

Nothing from this period remains on Earth. It is only dating of the chemical elements, the study of precious

■ APPEARING MINUTE IN THE PHOTOGRAPH ON THE OPPOSITE PAGE, EGYPT AND THE RED SEA ARE HERE CLEARLY VISIBLE TO ASTRONAUTS IN THE SPACE SHUTTLE. ON THE LEFT ARE THE NILE AND THE SUEZ CANAL; IN THE CENTRE, THE MOUNTAINS OF THE SINAI PENINSULA; AND ON THE RIGHT, SAUDI ARABIA AND, FARTHER UP AND TO THE RIGHT, ISRAEL AND THE DEAD SEA.

metals such as gold, platinum, and uranium, as well as comparative planetology, that enable us to reconstruct this Archaean era of the Earth's prehistory. Since then, the original impact craters and volcanic regions have disappeared, and the atmosphere and oceans have radically altered. More than any other body in the Solar System, the Earth has changed: quite literally, we are not living on the same planet.

From its very origin, our world has never stopped evolving. In what is perhaps a unique feature among the planets in our system, the terrestrial crust is mobile, consisting of rigid plates, which are constantly displaced by the underlying motion of the mantle. Once its tormented birth, hidden beneath a thick greyish veil, had passed, the atmosphere of the Earth cleared, and the Archaean history of the planet – interrupted by horrendous impacts – was able to begin.

The primordial ocean, dark and greenish in colour, lifeless, covered the whole of the early Earth. In places, the central peaks and the gigantic terraces around the largest basins were still visible. Then the first granite platforms became differentiated and rose up from the planetary crust, soon to be engulfed by the seas, which ebbed and flowed like gigantic (and almost living) planet-wide tides. With the passage of time, the seas invaded whole continents, and then retired during seemingly endless winters, which saw the extent of ice floes and glaciers increase immeasurably. The granite of the emergent land was gouged out into deep valleys by the rasping action of ice and rock, and dug still deeper when the ice receded and rivers transported the sediments deposited by the winds on the desert areas towards the sea. Borne along by their underlying plates, the original continents broke up, and disappeared into the depths of the mantle along the giant faults at the oceanic subduction zones, only to re-emerge elsewhere in a different form. At the end of the Precambrian, the first named continent (Pangaea) appeared. Then this broke up, setting its own progeny, Laurasia and

■ ON THE BORDERS OF ALGERIA, NIGER, AND MALI, THE FIELDS OF DUNES AND THE SANDSTONE HILLS OF THE SAHARA DESERT EXTEND AS FAR AS THE EYE CAN SEE. THESE DARK CLIFFS, WHICH ARE TODAY PARTIALLY SUBMERGED BENEATH WAVES OF SAND, ONCE DROPPED INTO THE WATERS OF A LARGE, CALM LAKE. FOR NEOLITHIC PEOPLE, THE SHORES OF THE LAKE AT OUAN RECHLA MUST HAVE BEEN A VERITABLE EDEN.

■ PROTECTED FOR 8000 YEARS FROM THE SAND-LADEN WINDS BY A ROCK SHELTER, THIS LARGE CARVING AT OUAN RECHLA DEMONSTRATES THE SENSITIVITY OF THE SAHARAN ARTISTS.

Gondwana adrift, which later gave rise to Eurasia, America, and Africa. Even later, fragments of the continental land masses, such as Madagascar, Australia, Antarctica, and India broke away and started their journeys across the oceans, while the long cordillera that borders the Americas, in a constant state of upheaval, saw its volcanoes repeatedly erupting, sowing destruction, blasting away ancient mountains, and suddenly building others. In the middle of the Pacific, invisible beneath the waves, another chain, this time submarine, revealed its presence by the incessant activity of its volcanoes, which tower 9000 m above the ocean floor in immense shield volcanoes, capped between eruptions by snow and ice. Farther away, the titanic shock between the Asian continent and India created the finest example of a system of folded mountain ranges – a form unique in the Solar System. The terrestrial crust, compressed between the two continental plates, initially resisted, but then folded, raising its sediments, gneiss and granite rocks, creating the ranges of the Pamir, the Karakoram, and the Himalayas. In the winking of a geological eye – in just 10, 20 or 30 million years – the mighty mountain barrier transformed the continent, interrupted the routes of the dominant winds, upset the climate; raised an immense, icy desert plateau, and fertilized a thousand valleys. Seemingly touching the sky, the nameless mountains, which would one day come to be known as Machhapuchhare, Annapurna, and Daulaghiri became covered in snow and ice. These immense summits, which seem the very image of power and of lasting permanence, are, however, some of the most fragile of nature's creations. Glaciers, waterfalls, and mountain torrents erode them, streams and silt-laden rivers sculpt them, and tirelessly transport alluvium to enrich the low-lying land of the deltas.

The creature slowly emerges from the muddy sea-floor in which it had buried itself, leaving only its antennae and two large, dull and expressionless eyes visible. The oval, armoured head, emerged, soon to be

■ The desert seen from the sky. In the central Sahara, the wind sculpts the vast expanses of sand into immense fields of dunes. Although the large animals of the Neolithic — the giraffes, elephants, rhinoceros, and big cats — have deserted this inhospitable country, migrating to the savannas and forests farther south, life still clings to this almost Martian landscape.

THE BIOSPHERE: PROTECTED
FROM METEORITES AND FROM THE
DEADLY SOLAR RADIATION BY A
THIN ATMOSPHERIC LAYER, LIFE
HAS CONQUERED EVERY NICHE
ON EARTH, FROM THE DEPTHS
OF THE OCEANS TO THE SNOW-
CLAD PEAKS. BUT IT HAS ALSO
PROFOUNDLY MODIFIED THE
PLANET. ATMOSPHERIC
OXYGEN HAS BEEN
PROVIDED IN
ABUNDANCE BY
PHOTOSYNTHETIC
PLANTS.

followed by the three-lobed, segmented body, armed with three pairs of spines. Nearly five metres down, the animal waves around between bushy colonies of graptolites and the rocks. Above it, not far from the surface, a translucent jelly-fish unconcernedly expands and contracts its canopy in the warm water. The oceans are completely empty except for this narrow, and extremely

■ An Amazonian mangrove swamp hides its billions of living inhabitants. Climbing perch that venture onto the branches; billions of insects which invade the foliage and mercilessly attack the rare tree species; mammals and reptiles which hunt and fight over their prey; frogs and birds that croak and shriek when night falls.

shallow, coastal zone. At the surface, there is nothing. A desert of greyish granite, of vast basalt plains dominated by the symmetrical cones of volcanoes, whose summits are enveloped by the hot, thick, heavy atmosphere, through which the moonlight filters with difficulty – moonlight from a Moon that is strangely close in the sky. If we did not know that this sea-floor is that of the Earth on a fine night, some 570 million years ago, we might well think that we were on another world. We are at the beginning of Earth's history, at the very dawn of the Palaeozoic era. Before that was prehistory: the Archaean, a lost era. The longest period on Earth left so little in the way tangible traces: no continent that we can name, unknown atmospheric conditions and weather, disrupted by cosmic catastrophes whose scale and consequences are quite unknown.

Life on this planet appears in the primordial ocean at the very dawn of the Archaean era, about 3.5 billion years ago. We have not the slightest idea of how the most complex molecules came together to form amino acids, proteins, and then the nucleic acids which would finally bring the first living being into existence, an extra-ordinary combination that was capable of surviving, of reproducing, and of adapting to its environment. The carbon atom, with its astonishing power to combine with other atoms, and water, the ideal medium – thanks to its solvent properties, and its high thermal inertia – allowed the construction of more-and-more complex chemical factories. Life is born of water: but did it arise on the clays along the shores of the very first

■ Life has existed on Earth for at least 3.5 billion years. Blindly following the most unexpected evolutionary paths, it has, over the course of time, developed increasingly strange and complex forms; here adapting itself to a new climate; and there evolving to cope with the shock caused by the collision of two different regions, driven together by continental drift.

continents, or in the boiling-hot waters of the deep-sea 'smokers', some 5000 m deep? We don't know.

For almost the whole of Earth's prehistory, bacteria and blue-green algae, the cyanobacteria, patiently built stromatolites (the first known fossil structures) at the bottom of the sea. Strange domes that rose from the sea-floor at depths of 15 metres or so, and occupying areas kilometres across, stromatolites were practically the sole life forms on our planet for 2 billion years. With the appearance of eucaryotic cells, and then sexual reproduction, the first animal forms arose, about one billion years ago. With its unknown geography, metamorphic rocks, extremely rare and poorly preserved fossils, the records of this era have practically all been destroyed. The Earth's history thus begins with the geological period known as the lower Cambrian, dated some 570 million years ago. Three billion years had passed since the origin of life.

To palaeontologists, this era is the richest, the most extraordinary, and also the most mysterious of the Earth's whole history. The sediments show a sudden profusion of fossils, in a striking state of preservation. It was at this period that life had the good sense to invent exoskeletons, shells, and tests, all of which are well preserved in fossiliferous strata. But, above all, starting in the Cambrian and over a few tens of millions of years – i.e., in just 1 per cent of its overall existence on Earth – life underwent what can only be described as an explosion, an incredible radiation in every possible direction. It was in the Cambrian that almost all of the great phyla appeared that would later give rise to all the forms of life that followed one another throughout the next 500 million years. Since that time, the major anatomical forms have been fixed. It is to *Pikaia gracilens*, the precursor of all fish, and to its successors that swam in the Silurian seas, that we owe our vertebral column (which we share with sharks, frogs, alligators, elephants, bullfinches, and numerous

■ THE BAHAMAS COMPRISE MORE
THAN SEVEN HUNDRED ISLANDS,
SCATTERED IN THE ATLANTIC OFF
FLORIDA, CUBA, AND PUERTO
RICO. IN THE FOREGROUND WE
CAN SEE THE RELATIVELY HIGH
GROUND OF THE ISLAND OF
ANDROS, WITH NASSAU AT ITS
LEFT END, SEPARATED BY THE
DEEP TONGUE OF OCEAN
CHANNEL. BEHIND THAT,
ARE THE NARROW CORAL
REEFS OF ELEUTHERA
AND EXUMA.

other animals). This exuberance of life, which would burst forth into cnidarians, annelids, molluscs, echinoderms, arthropods, chordates, more than ten thousand species of trilobites, and an astounding number of extravagant creatures that we will never know, for lack of fossil remains, is perhaps the result of the low selective pressure that prevailed in the Cambrian oceans, and then in the Ordovician, Silurian, and Devonian periods. At that time, the almost infinite possibilities offered by genetic mechanisms gave rise to every form that was capable of surviving and reproducing. 'Nature, which has no specific goal, tries everything', as the biologist Albert Jacquard has said. Natural selection, the mechanism behind the evolution of species that was discovered by Darwin, naturally operated then, but undoubtedly in a less rigorous fashion than in later periods.

During the Palaeozoic era, ecological niches were large, comfortable and often unexploited, and thus allowed the appearance of a myriad different species, both strange and familiar, with destinies that were to be either glorious or obscure, as is shown by *Pikaia gracilens* and *Olenellus nevadensis*. The first is an ancestor of the vertebrates, while the second has disappeared in the depths of time. The trilobite that we described earlier, emerging from the mud of the lower Cambrian ocean was *Olenellus*. Conventionally, it is this particular species that has the honour of marking the beginning of the Palaeozoic. Before *Olenellus* is prehistory; as soon as it appears in the strata, we have arrived at the modern era. From then on, everything accelerates: in the Devonian, some 400 million

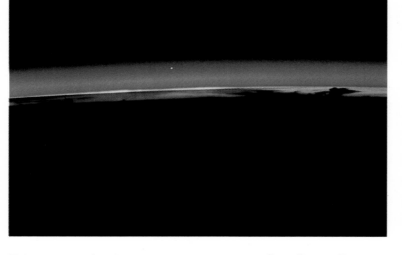

■ In 1989, the American astronauts on board the Space Shuttle Discovery launched the Magellan probe towards Venus. After one sunset, which occurs every hour in Earth orbit, they took this striking photograph of the Earth's sister planet emerging from the upper atmosphere.

years ago, a rather undistinguished crossopterygian fish – like modern-day dipnoids and coelacanths – left the water. It had strong short fins, bronchi, and rudimentary lungs, and was capable of moving awkwardly from one pool to another. When these dried up, it buried itself in the still soft mud, which then hardened around it. It could remain there in a state of suspended animation for months, or even years. If the water was not too late in returning, it emerged from its muddy shroud until the next dry period. Along with the first semi-terrestrial vertebrate species, perhaps *Eusthenopteron*, gymnosperm plants and arthropods slowly but surely invaded the shores of the continents.

In the Devonian, Carboniferous, and finally the Permian, our planet took on a more familiar appearance. Immense forests of tree-ferns sheltered a fauna consisting of insects and batrachians, often giant forms: the dragonfly *Meganeura* had a wing-span of no less than 70 cm; some millipedes, such as *Arthropleura*, reached a length of 2 m, and fought for prey with *Megarachne*, a spider 50 cm across. These nightmare creatures, however, already had their own predators, the very first reptiles, who, 350 million years ago, left the water and conquered all of Pangaea. At the end of the Permian, 240 million years ago, the Earth's future seemed to be settled: a planet with an eternally tropical climate, forever inhabited by insects and reptiles. The ancestors of sharks were already prowling Earth's unique ocean, as well as the trilobites, which, after 320 millions years of adaption, seemed to have exhausted their evolutionary potential and were returning to the simple forms of the Cambrian. They disappeared at the end of the Permian, and with them the Palaeozoic came to its conclusion. The transition from the Palaeozoic to the next era – the Mesozoic – is not just a convenient convention, but corresponds to a major ecological upheaval on our planet. In fact, at the end of the Palaeozoic more than 90 per cent of living species disappeared from the surface of the Earth. What caused this mass extinction? Was it through geological and climatic variations, such as mountain building or sea-

■ EVOLUTION AS SHOWN BY THREE TRILOBITES. FROM TOP TO BOTTOM, *OLENELLUS*, WHICH MARKS THE BEGINNING OF THE PALAEOZOIC; *MODOCIA*, A TYPICAL CAMBRIAN SPECIES; AND FINALLY *AULACOPLEURA*, WHICH LIVED 420 MILLION YEARS AGO. TRILOBITES AS A FAMILY BECAME EXTINCT AT THE END OF THE PALAEOZOIC.

level changes? Was the cause of the extinction external to the Earth, through the explosion of an exceptionally close supernova, or the impact of a large comet? We simply do not know.

When the Triassic period opens, we therefore find an almost deserted Earth awaiting the arrival of a new fauna. This is the start of the Mesozoic era, the famous era of the dinosaurs. The latter are, in fact, a particular class of reptiles, which are distinguished from archaic and modern reptiles primarily through their mode of locomotion. Dinosaurs were undoubtedly the dominant vertebrates during the Mesozoic era. First appearing during the Triassic, they gradually occupied every ecological niche during the Jurassic and the Cretaceous. On Gondwana and Laurasia, and then in Africa, America, and Eurasia, quadrupeds, bipeds, herbivores living in herds, and solitary carnivores, all attained colossal sizes – 15 m high, 30 m long, and 70 tonnes, or more, for *Diplodocus*, *Brachiosaurus* and *Ultrasaurus* – or, in contrast, were no larger than a pigeon. In the oceans, the plesiosaurs and ichthyosaurs became so well adapted that their shapes sometimes closely resembled those of fish and dolphins. In the air, pterosaurs were able to glide for hours, or even days, and also attained incredible sizes: the membranous wings of *Pteranodon* and *Quetzalcoatlus* had spans of 8–10 m. Did these magnificent long-distance fliers notice, some 150 million years ago, the appearance of a mutant among the reptiles, a bizarre creature, that would subsequently enjoy great evolutionary success? This was *Archaeopteryx*, which had two wings that were covered in feathers and carried claws, a long reptilian tail, and a mouth armed with teeth! This was the first bird – or at least one of the first. The size of a chicken, did this animal hunt, among the roots of the conifers, for even smaller creatures with cheeky snouts covered in fine whiskers that were the first mammals, and which had already been living in the shadow of the great predators for some 50 million years? The tiny *Megazostrodon* was probably nocturnal, catching insects and gnawing at the carcasses of dead

■ GRAND CANYON, CUT BY THE
COLORADO RIVER IN ARIZONA,
ALLOWS GEOLOGISTS TO EXAMINE A
LARGE PORTION OF THE HISTORY OF
OUR PLANET. ITS COLOURED
SEDIMENTARY LAYERS, SLOWLY
EXPOSED BY THE RIVER'S
EROSION, HOLD FOSSILS THAT
BEAR WITNESS TO SEVERAL
HUNDRED MILLION YEARS OF
EVOLUTION, FROM THE
PALAEOZOIC TO THE MOST
RECENT ICE AGES.

■ THE EARTH AND MOON FORM
AN UNUSUAL PLANETARY PAIR,
UNIQUE IN THE SOLAR SYSTEM. THE
GRAVITATIONAL LINK BETWEEN THE
TWO BODIES HAS POSSIBLY
STABILIZED THE EARTH'S ORBIT,
AND THUS ALLOWED LIFE TO
DEVELOP ON OUR PLANET. THE
EFFECTS OF THE LUNAR TIDES
ON THE OCEANS MAY NOT BE
UNRELATED TO THE
APPEARANCE OF LIFE ON
EARTH.

dinosaurs that lay, like beached ships, on the shores of Mesozoic lagoons.

Unlike the end of the other known geological periods, that of the Mesozoic era is particulary sharp. In sediments dating from around 65 million years ago, all over the Earth there is a thin layer of limestone, less than 10 cm thick, which marks the boundary between two different worlds. Below it, we have the end of the Cretaceous, when the world was still ruled by powerful *Triceratops* and, above all, by *Tyrannosaurus rex,* the most powerful terrestrial predator that has ever lived. Above this layer of limestone, we have the Eocene, the first period in the Tertiary, or Caenozoic, era. By the Eocene, all the dinosaurs have disappeared, as well as 75 per cent of all the species living in the Cretaceous. Nothing would be the same again. In the Tertiary, mammals, birds, and flowering plants would undergo explosive evolution. The modern-day Earth, with its five major continents and its various large islands, would finally take shape, and the great, modern, folded mountain chains of the Alps, the cordillera of the Andes, and the Himalayas would arise, while a particular family of primates would evolve and stand upright. What happened during that extremely short lapse of time that separates the two eras that are so radically different? After several decades of conjecture, some of it quite unrealistic, one fairly robust and well-supported scenario has emerged, with fairly unanimous acceptance: that 65 million years ago, an asteroid or a comet crashed onto Earth.

The impact – the site of which has probably been found recently off Chicxulub, in the Gulf of Mexico – would have disrupted our planet's delicate atmospheric equilibrium. Creating a mountain of water, 1000 m high, which would devastate the American coast, the impact would have ejected some tens of cubic kilometres of dust into the atmosphere, darkening the sky for several years. Deprived of heat and light, the Earth's surface would have been plunged into a long winter night that would have been fatal for most species of plants and animals. The average temperature at the surface of the ground would have dropped from +20 °C to − 10 °C, and the oceans would have frozen. For unknown reasons, birds and mammals survived this decimation, and subsequently conquered the planet. The tiny and unassuming, but still numerous and hardy, arthropods and insects also survived these great geological upheavals unaffected.

About 2 million years ago, *Homo habilis* appeared among the placental mammals that lived on the African savanna, the descendant of a long line of primates dating from the end of the Mesozoic. Unlike its cousins, the great apes of the Pliocene, and rather better than its ancestor, *Australopithecus afarensis, Homo habilis* used tools and weapons made from pebbles and the branches of trees. A descendant, *Homo erectus* tamed fire and invented language. Slowly colonizing the whole of Africa and Asia, the species would finally give rise to *Homo sapiens* a few hundred thousand years ago, at the beginning of the Pleistocene. This was the time of the great glaciations that came at the start of the Quaternary era. Some 50 000 years ago, *Homo sapiens* hunted mammoth and woolly rhinoceros, cave bears, and sabre-toothed cats. These humans dressed in skins, lived in communities, initiated the cult of the dead, and dug the first graves. They wore ivory jewellery and devised a new form of relationship with the natural world around them. The first religious rites accompanied their hunts, the seasons took on sacred significance, eclipses were regarded as signs, stars (and particularly planets) were regarded as having an influence on human affairs. Hidden in the caves at Lascaux, or on the shores of the lake at Ouan Rechla in the Sahara, *Homo sapiens* painted and engraved pictures on the sandstone walls. Conscious of their own existence, they discovered the anguish of death and expressed it through sublime art. Without knowing it, they had become capable of conscious thought: *Homo sapiens sapiens*.

In 4.55 billion years of evolution, the Earth has changed. A living planet because of its extreme geological activity, it is also one thanks to the complex and subtle relationships found in the life-forms that it has developed. Life has profoundly altered the terrestrial atmosphere. The primitive blue-green algae, which for hundreds of millennia constructed stromatolites, also enriched the air with oxygen. Their photosynthesis would eventually lead to the day when *Eusthenopteron* would be able to leave the water. Once rich in carbon dioxide, like the atmospheres of Venus and Mars, Earth's atmosphere now contains less than one per cent. This greenhouse gas is essentially trapped in the oceans, carbonate rocks, and plants, which use it to grow, emitting oxygen as they do so. The result of this subtle relationship between the atmosphere, the Earth, and vegetation speaks for itself. The average temperature of the surface of our planet is about $+15\,°C$. Without any atmosphere at all, the thermometer would read about $-25\,°C$. On Venus, the temperatures exceed $450\,°C$ and on Mars they are about $-60\,°C$. Under terrestrial conditions, and uniquely in the whole Solar System, the hydrological cycle is permanently active, practically every-

■ PHOTOGRAPHED FROM THE SHORE OF LAGUNA VERDE, THE SNOW-CAPPED 6900-M PEAK OF THE VOLCANO NEVADO OJOS DEL SALADO DOMINATES THE ATACAMA DESERT. IN PLACES, THE FREEZING COLD WATERS OF THE SALT LAKE, WHICH IS SOMETIMES VISITED BY OCCASIONAL PINK FLAMINGOES, ARE HEATED BY BOILING-HOT SPRINGS.

where on the planet. Heated by solar radiation, the water from the oceans evaporates into the atmosphere, which transports it, through the general west–east circulation, towards the land. There, clouds are cooled and the water vapour condenses into ice crystals or drops of rain. The Earth's gravity causes the water to fall as snow or rain, where it feeds the soil and the living creatures that inhabit it. Irrigating the plains, and eroding arid plateaux, water, initially in the form of glaciers or waterfalls, returns through streams and rivers to the sea.

Life shapes terrestrial landscapes. The high chalk cliffs that edge the coast consist of billions of fossil organisms. The regular, purple and gold patchwork of the lavender and wheat fields of Provence, the limitless prairies of Montana, the impenetrable Amazonian rainforest all form an even layer of humus and soil that covers practically the whole surface of the continents. A layer of living matter… What is more, the overall symbiosis that links life to the substratum from which it sprang has caused some people to adopt the view that the Earth itself, Gaia, is alive.

Although the overall properties of the Earth's atmosphere are determined by great astronomical cycles, its different climates are determined by local topography. In South America, the two coasts – Atlantic and Pacific – are radically different. In the east there are vast primitive forests that are perpetually humid. In the west there is a desert coast, where, in parts of the Atacama Desert, rain never falls. Icy waters brought from the Antarctic by the Humboldt Current flow along the Chilean coastline. The cold Pacific waters do not evaporate, and just a few metres from the ragged coastline that is battered by the waves, there is an absolute desert. In Asia, it is the Himalayas that create both rain and fine weather. Every year, when summer arrives, the gigantic wall acts as a barrier to the clouds from the Indian Ocean, and initiates the monsoon, which fertilizes millions of square kilometres of land, but which also sows death when torrential rains arrive in a cyclone. This water does not cross the mountain chain, and to the north of the Himalayan peaks, there lie desert steppes, plateaux where life has difficulty surviving. The region around Leh, the Takla Makan, and the Gobi Desert will have to wait millions of years for regular rains.

As for the story of Europe, it has been played out beneath the influence of

extremely strong sunlight – in the Gulf of Mexico! The warm current of the Gulf Stream crosses the Atlantic and brings changeable weather everywhere from the valley of the Guadalquivir in Spain to Flanders, and even as far north as Spitzbergen, affecting Perigord and Cornwall in between – over an infinite variety of landscapes with constantly changing seasons.

Geological upheavals, the slow drift of continents, and the growth of mountains, in changing the climate have also caused changes in living species. In the heart of the Sahara, where, lost in the waves of sand that form the dunes, a few acacias fight desperately against drought, scorpions and gerbils hide in caves with decorated walls. Ten thousand years ago, *Homo sapiens* lived here, hunting antelope and giraffe. In the Cambrian, and at precisely the same spot, *Olenellus* emerged from the ooze … When there is a warm, wet period, there is an explosion of plants, accompanied by billions of insects. When drought arrives, reptiles invade the Earth. When two continents that have been distinct for a vast period of time collide, there is a major ecological upheaval. In the Tertiary, the two Americas became joined by the Isthmus of Panama. To the south, there were marsupial mammals; to the north, placental mammals. There was a confrontation between two of nature's different experiments, with the results that we see today. The infinite abundance of genetics has ensured, over 3.5 billion years, that life has never disappeared completely, even when there have been major ecological crises. By creating random mutations, whose viability is by no means certain, nature has blindly followed the unexpected twists of fate. Sometimes, however, it gives up. In this respect, the Antarctic continent, carried across the Indian Ocean towards the South Pole has been unique. Its climate slowly altered, until it became ultra-continental – suffering from extreme drought and cold, with no possible end in sight. Life, here, did not pick up the gauntlet: its fauna and flora are dead, buried beneath kilometres of ice. Today, only a few colonies of birds cling to its coasts – and a few tens of *Homo sapiens sapiens*.

Human beings, like *Archaeopteryx* or *Triceratops,* are an accidental result of the Earth's history, and not an inescapable consequence of the laws of physics. The fate of stars and

galaxies was decided at their birth, and we can predict when and how they will be born, will evolve, and will fade away. The story of the Earth is unique, and its future is unpredictable. The chaotic behaviour of plate tectonics and of the mechanisms that determine climate; life's chance fall of the dice; the interplay of natural selection and adaptation; and the potentially fatal dance of the asteroids above our heads, all mean that the range of possibilities is wide open.

Nowadays, although only seemingly, human beings have changed the rules of the game. By transforming the landscape – over a long period of time, and quite unwittingly – by decimating the flora and fauna, and disturbing the ecology, our species has introduced insidious, and perhaps deep and lasting, changes to the climate. We may be masters – perhaps unreasoning ones – of the planet, but we are not masters of its fate. No one today can predict the exponential growth of a population, whose energy expenditure is thousands, or even millions of times greater than that of all the other species combined. But lurking in the shadows, one of nature's great success stories, the arthropods, resist our industries, our chemical warfare, nuclear accidents, and pollution. Consisting of a million species – and thus billions and billions of individuals blessed with an astonishing reproduction rate, offering immense genetic variability – the order of insects alone has an almost infinite potential for developing future species. So, if, despite everything, we had to try to give a definite prediction of the future of the Earth, we might evoke the scene one early morning in the year of grace 5 000 000 000. In the east, on a planet still populated by insects, but where all memories of humanity have long since been erased, the immense, red, turbulent sphere of the Sun rises for the last time, filling the horizon, before snuffing out our tiny planet and its satellite in a flash, as it sheds into space the beautiful, electric-blue shell of a planetary nebula.

■ STROMATOLITES ARE THE OLDEST LIVING FORMS KNOWN ON EARTH. VERY RARE COLONIES OF THE MINUTE BLUE-GREEN ALGAE HAVE SURVIVED THE AGES DOWN TO OUR OWN TIME. AS WITNESS, HERE IS A BED OF STROMATOLITES, LIVING ON THE SHORE OF THE INDIAN OCEAN, AT SHARK BAY IN AUSTRALIA.

■ A CRESCENT EARTH. THE TERMINATOR IS THE LINE SEPARATING LIGHT FROM DARK, DAY AND NIGHT, WHICH PERMANENTLY ENCIRCLES OUR PLANET. HERE, SOUTH AMERICA AT DAWN IS LARGELY ENSHROUDED IN CLOUD, AND THE SUN IS ABOUT TO RISE OVER EASTER ISLAND, BEFORE ILLUMINATING THE EMPTY WASTES OF THE SOUTH PACIFIC.

The Moon:

setting foot on another world

■ A DESERT LANDSCAPE, COVERED IN ANCIENT, SOLIDIFIED
LAVA SEAS, PEPPERED WITH BILLIONS OF METEORITIC
IMPACTS: THIS IS THE MOON AS IT APPEARED THROUGH
THE PORTHOLE OF AN APOLLO COMMAND MODULE.
THIS PHOTOGRAPH SHOWS THE REGION OF SINUS
MEDII WITH, IN THE FOREGROUND, THE CRATER
TRIESNECKER AND ITS FAMOUS RILLES. IN THE
BACKGROUND IS THE CHAIN OF CRATERLETS
FORMING THE HYGINUS RILLE, ANOTHER
LUNAR FORMATION CLEARLY VISIBLE
WITH A TELESCOPE FROM EARTH.

■ THE EXPLORATION OF OTHER WORLDS HAS ALREADY BEGUN. MORE THAN 3 BILLION YEARS AFTER ITS APPEARANCE ON EARTH, LIFE HAS ESCAPED FROM ITS PARENT PLANET: TWELVE MEN HAVE WALKED ON THE MOON.

An ashen landscape stretches as far as the eye can see. A sort of grey sand covers everything; the deserted plain on which we stand, as well as the mountains whose gentle contours and uncertain heights surround us on the horizon. A few dark basaltic rocks, half-buried, lie here and there on the ground. It is a landscape where there is no sense of scale and no landmarks: there is nothing here that helps to stop you from feeling disorientated. The total absence of any atmosphere causes objects to appear strangely close, and the world's strong curvature limits your field of view. In the velvet-black sky, a few stars struggle to shine, but are utterly drowned out by a harsh, immobile Sun. Worn-down impact craters break the ashen expanse. This utterly bare spot seems almost artificial. With its unmoving stars in a cloudless sky, this plaster landscape, bounded by an abnormally close horizon, and covered in footprints, seems like the abandoned set for a science-fiction film.

Yet there is one element in the landscape that is disconcerting, but real, and even moving: that droplet of water that hangs over the dreary hills, a brilliant blue ball that certainly appears to be alive. The white clouds that trail across it, hiding its oceans, and hugging its mountain chains, slowly change and disperse. Nothing on the world at our feet looks anything like that blue ball, with its beauty and poignant fragility. A living planet hangs in the sky of a world that is completely dead.

We are on the Moon, some 380 000 km from Earth. It is a lifeless body, with no seasons, and no atmosphere. Rocks that are subjected to the radiation from the Sun reach temperatures of 125 °C. A few tens of metres away, the dust hidden in the shadows of the mountains is at the frigid temperature of −175 °C. The thick layer of rather grainy dust that uniformly covers the lunar surface, here, at the base of the Apennine Mountains, has been blown away and churned up. The grey soil is covered in footprints. Someone has walked on this dead world. Some living being, a biped, has walked across these valleys, these craters, and these mountains. Some stranger to this world has been walking here. Science fiction?

No. From their home on the neighbouring planet, twelve men have indeed walked on the Moon. A short hop in terms of the Solar System, but a giant step for the human race, which, for the first time in its long history, has been able to escape its

■ The Earth's satellite is as large as a planet, but is too low in mass and too close to the Sun to have been able to retain its primordial atmosphere. Certain regions of the Moon, which suffers practically no erosion, retain the mark of impacts that are more than 4 billion years old.

terrestrial existence, moving beyond the boundaries of its own planet. After its appearance on Earth, some 3.8 billion years ago, and after a long and tumultuous history, life has managed to escape from its cradle and set out to discover the universe. These extraterrestrial expeditions took place between 1969 and 1972, in six successive stages. Too complex and too costly, they did not result in the establishment of a scientific base or even the colonization of the Moon, which has not felt the tread of human feet since 14 December 1972, nearly thirty years ago.

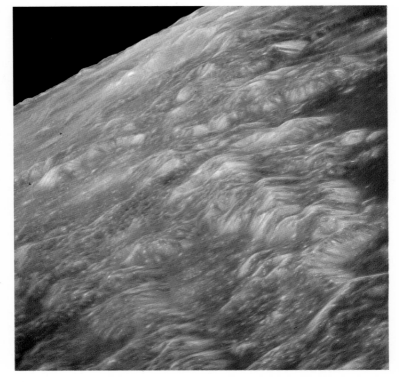

■ LUNAR MOUNTAINS HAVE VERY GENTLE SLOPES BUT MAY REACH HEIGHTS OF SEVERAL KILOMETRES. IN THIS PHOTOGRAPH, OBTAINED FROM LUNAR ORBIT BY THE APOLLO 17 ASTRONAUTS, THE RELIEF IS INDISTINCT BECAUSE OF THE LACK OF SHADOWS. CLOSE EXAMINATION, HOWEVER, REVEALS THE MARKS LEFT BY THOUSANDS OF IMPACT CRATERS.

The Moon is the sole natural satellite of the Earth. Measuring 3486 km in diameter, it completes one orbit, a circuit of more than 1.2 million kilometres, around our planet in slightly more than twenty-seven days. At the closest point in its orbit, at perigee, it lies at a distance of 353 880 km from the centre of the Earth. At its maximum distance, at apogee, it is 421 690 km away. Much lighter than the terrestrial planets, with a density of 3.34 as against 5.45 for Mercury, 5.25 for Venus, and 5.52 for Earth, its mass is about 73 billion billion tonnes: 1/81 of that of the Earth. Because of its mass and its physical and chemical characteristics, the Moon must be considered as a world in its own right. In size, it is the thirteenth body in the Solar System. Almost five times as massive as Pluto, it is only about 22 per cent of the mass of Mercury. Finally, the Moon is comparable with the other giant satellites in the Solar System: Io, Europa, Ganymede, Callisto, Titan, and Triton, which

orbit Jupiter, Saturn, and Neptune. Their diameters (between 3640 km and 5280 km) are close to, and sometimes exceed that of the Moon. Their densities (between 1.80 and 3.53) are, in most cases, less and, finally, their masses are comparable (between 48.6 and 149 billion billion tonnes).

Yet the Moon is a unique body in the Solar System. Unlike the other large satellites, it is essentially rocky, without any ice, neither on its surface, nor, apparently, in the sub-surface layers. Finally, and above all, together with the Earth it forms a planetary pair that is unique in the Solar System (apart from the very special case of Pluto and Charon). All the large satellites just mentioned are, in fact, governed by planets that are between 500 and 40000 times their mass. When compared with Jupiter or Saturn, Io or Titan are tiny pebbles. Their gravitational influence on their parent planet is practically zero. The Moon, on the other hand, is just 3.6 times smaller, 81 times less massive than the Earth, but possesses 83 per cent of the system's angular momentum. Rather than being viewed as a satellite orbiting a planet, the Earth-Moon system should be seen as a double planet.

Because the Moon is so close to the Earth, it is strongly influenced by the gravitational field of its massive neighbour. Its rotation, for example, has been braked by the strong gravitational force, and has subsequently become precisely synchronized with its orbital revolution. As a result, it now permanently turns the same face towards the Earth. In return, it exerts a gravi-

■ RENDEZVOUS IN LUNAR ORBIT. THROUGH A PORTHOLE IN HIS SPACE CAPSULE, THOMAS MATTINGLEY PHOTOGRAPHED HIS TWO COMPANIONS, WHO, FOR THREE DAYS, HAD THE CHANCE TO SURVEY THE HIGH PLAINS NEAR THE CRATER DESCARTES. JOHN YOUNG AND CHARLES DUKE WERE PREPARING TO DOCK THEIR LUNAR MODULE WITH THE COMMAND AND SERVICE MODULES WHICH WOULD RETURN THEM TO EARTH.

tational attraction on our planet that is by no mean negligible, and quite literally displaces the water in the oceans by creating tides. The gravitational interaction between the two bodies creates extremely long-term effects on both. For example, the constant movement of billions upon billions of tonnes of water in the oceanic tides acts like mechanical friction and causes a loss of energy from our planet. The latter's rotation on its axis is imperceptibly slowing down. By studying the growth lines of corals and other fossil shells that date back to the Palaeozoic, palaeontologists have confirmed this slowing-down, and have discovered that the length of the day, early in the Palaeozoic, some 600 million years ago, was less than

twenty-two hours. As for the terrestrial year, it then consisted of more than four hundred days. The progressive dissipation of energy by the tides also produces a progressive increase in the distance of the Moon. Our satellite is currently receding from us at the rate of a few centimetres per year. At the dawn of the Palaeozoic, the first living creatures that ventured out of the water to conquer dry land had, in their sky, a Moon that was half the distance that it is today.

It was towards the end of the afternoon of 30 July 1971, at the foot of the lunar Apennines, at 3°39′E, 26°6′N. The astronauts were getting ready to leave their descent module to visit the area around the great mountain chain. It was by far the

■ IN DECEMBER 1972, ABOUT THIRTY YEARS AGO, THE STRANGE SILVERY AND GOLDEN BEETLE FROM EARTH LANDED ON THE MOON FOR THE LAST TIME. THE LUNAR MODULE REMAINED FOR SLIGHTLY MORE THAN THREE DAYS IN TAURUS LITTROW VALLEY, ON THE BORDERS OF MARE SERENITATIS, AT LONGITUDE 30° 45' EAST, LATITUDE 20° 10' NORTH.

most spectacular of all the lunar missions. For the first time, in fact, the astronauts had set their module down in an extremely rugged region. It was a chance for David Scott and James Irwin to admire the mountains, to visit the sinuous track of an ancient lava flow, and to drive their rover for nearly 28 km across the grey soil. Before them, Neil Armstrong and Edwin Aldrin were the pioneers who, on 20 July 1969, had taken their historic first steps around their Apollo 11 lunar module, on the monotonous desert of Mare Tranquilitatis, before taking off again, twenty-one hours later, to return to Earth. Then in November 1969, Alan Bean and Charles Conrad landed their module in Oceanus Procellarum, a few tens of metres from a space-

probe, Surveyor 3 that had been sent to the Moon two years before. After the failure of the Apollo 13 mission, Alan Shephard and Edgar Mitchell had to wait until 5 February 1971 to visit a new site, not far from the crater Fra Mauro, and only 100 km to the east of Apollo 12.

At their site in the lunar Apennines, David Scott and James Irwin enjoyed several precious moments, such as when they stopped the rover alongside the small crater Saint George to pick up a few samples. Because the Moon has not the slightest trace of atmosphere, the astronauts had taken complex and heavy survival equipment with them. Their spacesuits were heated and pressurized, and supplied them with terrestrial oxygen, stored in a bottle that they carried on their backs.

Strong, not very pliable, but extraordinarily tough, they were also very heavy. Their helmets were fitted with visors coated with pure gold, designed to protect their eyes from the unbearable solar radiation. Thus equipped, during training on Earth the astronauts weighed some 200 kg and were incapable of the slightest movement. Luckily, weight on the Moon is one sixth of that on Earth, and there James Irwin and David Scott weighed no more than just 35 kg. Although their weight had decreased by a considerable amount, their mass, by contrast, had remained exactly the same. Because of this, and hampered by an enormous amount of inertia, they had to relearn how to walk. Because their spacesuits did not allow particularly precise movements, their gait was rather lumbering and cautious. To move more rapidly, Irwin and Scott progressed with small jumps – a bit like kangaroos. Their principal problem remained – just like a large vessel entering port – in coming to a stop or changing direction. The slightest stumble across a stone, and the mass of their spacesuits inevitably condemned them to falling flat on their faces.

After the amazing landscape of the Apennines, in April 1972 the Apollo 16 astronauts visited the Descartes highlands. Charles Duke and John Young, who would pilot the Space Shuttle ten years later, landed on a plateau lying some 7800 m above the average level of the lunar mare areas. Then, because

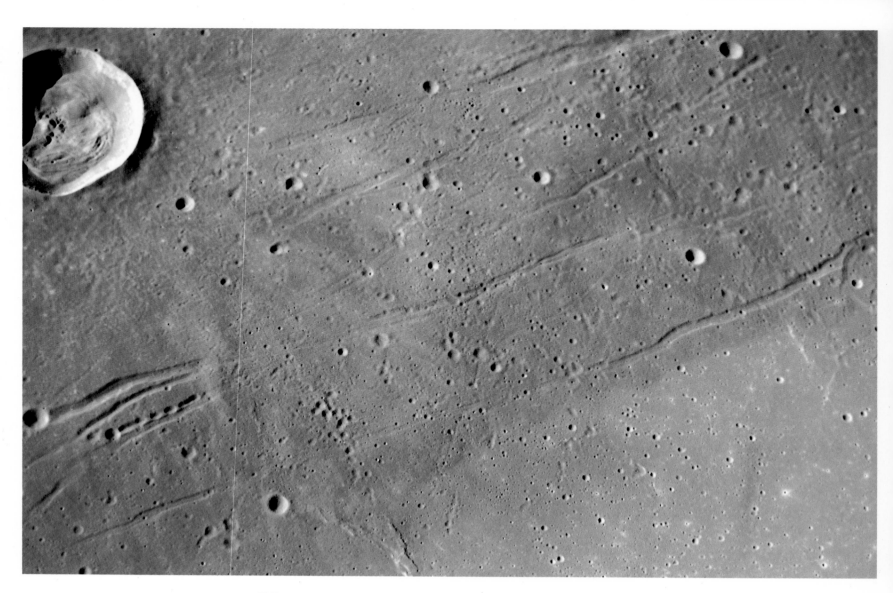

the Apollo 18, 19 and 20 lunar missions were cancelled for lack of funds and interest on the part of the public, the Apollo 17 astronauts brought the lunar adventure to a close in December 1972.

In the Apennines, that evening, well after the departure of Apollo 15 for Earth, the Sun takes several tens of minutes to set – after all, the lunar day lasts almost as long as a month on Earth. High in the sky, the Earth illuminates the landscape with an unreal, blue light. It never really gets dark: the blue planet, nearly four times as large as the Moon seen from Earth, provides a bright illumination, ten times as strong as the Full Moon that we are used to seeing. The stars move imperceptibly across the sky, but our planet remains almost fixed in the same place, unaffected by the slow rocking of the celestial sphere. The rotation of the Moon on its axis is tightly locked to its revolution around the Earth. Seen from the latter, only one hemisphere of the Moon is ever visible and, of course, the Earth is never seen from the hidden face of the Moon, which is the only region of our satellite where the night becomes truly dark after sunset. This hidden face of the Moon is, incidentally, coveted by modern astronomers, who see it as the ideal site for an observatory, and perhaps the best site for astronomical observation in the whole of the Solar System.

The Apennines form the most magnificent mountain system on the Moon. Consisting of several thousand individual peaks, they form a circular arc some 650 km long. The view, in the region of the Apennines, resembles that of the Andean Altiplano, or even some landscapes on Hawaii, or in the Massif Central. The peaks have a very subdued relief, with gentle slopes, and rounded hills, rather like very old terrestrial volcanoes. Here, at the Apollo 15 landing site, we are on the flank of Hadley Mons, which is as high as Mont Blanc (4800 m). A few tens of kilometres to the south rises Huygens Mons, which exceeds 5500 m in height. At the Moon's south pole, the last *terra incognita* on the Moon, the Doerfel and Leibniz mountains reach altitudes of perhaps 9000 m. These are the highest mountains on the Moon. The lunar peaks and the great mountain ranges have not been created, as they have on Venus, Earth, or Mars, by tectonic movements or 'hot spots'. The Apennines, like the Alps, which succeed them in the north, actually represent the remnants of an ancient impact. After its formation – which, as we shall see, is one of the greatest mysteries in the Solar System – the Moon underwent the intense meteoritic bombardment that was then affecting all the planets. Its surface was undoubtedly molten, as a result of the heat released in the course of its accretion and also from short-lived radioactive elements. The Moon subsequently cooled slowly, while a solid crust formed on its surface, and its primordial atmosphere escaped into space. The lunar crust, consisting of low-density basaltic and anorthositic rocks, has an average thickness of 70 km, covering a denser and more rigid mantle. We still do not know if our satellite has a central metallic core, like Mercury, Venus, Earth, and Mars.

■ BETWEEN A MARE AND THE LUNAR HIGHLANDS, A VERY DEEP CRATER, SEVERAL KILOMETRES IN DIAMETER, IS STILL ENVELOPED IN NIGHT-TIME SHADOW. TO ITS RIGHT, A MINUTE CRATER, WITH CLEAR-CUT EDGES, LIES IN THE CENTRE OF A VERY BRIGHT HALO. THIS IS AN IMPACT THAT DATES BACK SEVERAL TENS OF MILLIONS OF YEARS, AT THE VERY LEAST. AN ANCIENT LAVA FLOW APPEARS JUST ABOVE THIS CRATER.

■ IN THE GREY DESERT THAT BORDERS ON MARE SERENITATIS, ON AN AIRLESS WORLD, WHERE LIFE NEVER DEVELOPED, A VISITOR FROM ELSEWHERE MAY BE FOUND. ONE DAY, PERHAPS, OTHERS LIKE HIM WILL RETURN HERE TO INSTALL GIANT TELESCOPES BENEATH THE PERPETUALLY DARK SKY, AND CAPABLE OF STUDYING THE FARTHEST REACHES OF THE UNIVERSE.

Hardly had its crust cooled and hardened, some 4.4 billion years ago, when the surface of the Moon was saturated and constantly reworked by meteoritic bombardment. Some 4 billion years ago, the most important and last of these, which were undoubtedly caused by asteroids several tens of kilometres in diameter, excavated the great basins that remain visible today: Mare Crisium, Mare Serenitatis and, above all, Mare Imbrium, the largest of all, which measures 1200 km in diameter. It was the immense shock from the Mare Imbrium

■ HARRISON SCHMIDT HAS PARKED HIS ROVER ON THE GENTLE SLOPES OF A MOUNTAIN. ALL AROUND, THE LUNAR SOIL IS COVERED IN ROCKS OF ALL SIZES, WITNESS TO AN ANCIENT IMPACT THAT THREW THE DEBRIS OVER THIS TAURUS-LITTROW SITE. ON THE FRONT OF THE VEHICLE, LIKE AN UMBRELLA POINTING TOWARDS THE SKY, IS THE COMMUNICATION ANTENNA THAT KEPT THE ASTRONAUT IN TOUCH WITH THE EARTH.

impact that created the mountain chains of the Alps and the Apennines, raising ramparts that were 5000 or 6000 m high. These exceptionally long-lived mountains are still there, but the impact basin itself was completely flooded, about 3.8 billion years ago, by a veritable sea of lava that arose from deep within the Moon, where it was created by the transient radioactive heating of the rocks of the mantle. The planets, even the smallest of them, are bodies that evolve and which are geologically 'alive' for longer or shorter periods of time. It was this spectacular episode that saw the formation of the lava 'seas' (the maria) that are one of the characteristic features of the Earth's satellite, and which punctuate the alabaster surface to produce the 'Man in the Moon' visible from Earth, a white-faced clown, now sad, now questioning, and now pensive.

This ocean of magma finally covered all the lower areas of the Moon, amounting to some 20 per cent of the overall surface, and then slowly solidified.

In places, small volcanic domes, between 10 and 25 km in diameter and a few hundred metres high, tipped with minute calderas, bear witness to past volcanic activity on our satellite, long since ceased. Elsewhere, the sinuous courses of ancient rivers of lava remain visible. In places, the seas of lava from the Moon's Archaean era partially flooded even more ancient formations, which emerge oddly from the dark plains. Here, the ancient central peak of a crater, 1000 m high, stands, isolated, in a sea of

■ LEAVING ONE LAST FOOTPRINT IN THE LUNAR SOIL, THE APOLLO 17 ASTRONAUTS LEFT FOR EARTH. THIS TRACE OF THE PRESENCE OF A LIVING SPECIES ON THE MOON WILL PERSIST FOR MILLIONS OF YEARS.

frozen lava. There, the ramparts of ancient impacts form perfect circles in Oceanus Procellarum. The basaltic maria, dark, flat, and even, are covered in tiny impacts, which obviously occurred later. They bear witness to perpetual meteoritic activity in the Solar System, but also to the decreasing frequency of impacts, as well as a constant decrease in the size of the impacting bodies. Large craters are rare in the lunar maria, and most were formed 2.8–3.8 billion years ago.

The last cataclysmic impacts on the Moon, which created the Kepler, Tycho, and Copernicus craters, date back about one billion years. Copernicus is a magnificent ringed plain 90 km in diameter, the central peak of which reaches a height of 1200 m. Its terraced walls rise to heights of 3800 m. Proof of its recent formation may be seen in the ejecta that were deposited on Oceanus Procellarum at the time of the impact, and which remain perfectly visible, covering the dark lava with a fine deposit of white dust. With the exception of these last major impacts, the Moon today wears exactly the same face as it had 3 billion years ago: it is a world that is geologically dead.

Unlike the maria, which are young, low areas, covered in dark materials, the high, bright old regions, which occupy 80 per cent of the lunar surface, are known as highlands. They consist of an utterly chaotic jumble of overlapping and ruined impact craters, superimposed on one another to such an extent that they have become unrecognizable. The most ancient of these terrains, more than 4 billion years old, now appear as formless, jumbled, mountainous masses of material. The lunar highlands are witness to the turbulent origins of the Solar System. The Moon, Mercury (and perhaps Callisto) are the only bodies that have preserved the most ancient traces of our origin.

It is 14 December 1972: Eugene Cernan and Harrison Schmidt are the last men to walk on the Moon. They slowly climb back up the metal ladder leading to the

airlock of the Apollo 17 lunar module. Around them lies the valley of Taurus-Littrow, on the borders of Mare Serenitatis, at 30° 45′ E and 20° 10′ N. Harrison Schmidt is a geologist: the first and last scientific representative on the Moon, he carries all the hopes of the planetologists who yearn to understand the origin of our satellite. Apollo 17 stayed for slightly more than three days in the Taurus-Littrow valley, and the American geologist travelled some 34 km across the lunar surface on board the electrically driven lunar rover. During his return journey to Earth, he asked himself: would the rock samples that he had meticulously collected lift the veil over the birth of the Earth's satellite?

Thirty years later, and despite the 25 billion dollars spent sending twelve men to the Moon, who brought 385 kg of lunar rocks back to Earth, we know that the answer is 'No'. Although the evolution of the Moon over a period of nearly 4.5 billion years is well understood, no one yet knows where and how this small world was formed. Three hypotheses are still in contention. First, the double planet: Earth and Moon formed together by accretion at the very beginning of the Solar System. This appealing theory does not stand up well to scrutiny: according to the majority of planetologists, the two bodies differ too much in their chemical composition for them to have been formed from the same interplanetary mix. The second model envisages that the Moon was captured by the Earth. Some 4.5 billion years ago, the orbit of the Moon crossed that of the Earth, and the latter altered its path, and finally captured it. Here again, physical and chemical arguments and also mechanical considerations prompt most planetologists to reject this theory. According to experts in celestial mechanics, for a body with the mass of the Moon to be captured by one that is only eighty-one times as massive, the orbits of the two bodies would have to be extremely close from the very start. That then comes back to the problem just mentioned of the difference in the chemical composition of the Earth and the Moon.

The last hypothesis is the most spectacular, and is also the one most seriously considered by planetologists. In the Solar

System's very early stages, the Earth, which was still in the process of formation, was struck by another, smaller protoplanet, with a mass roughly equal to that of Mars. This impact, which was obviously the largest ever to have affected our planet, destroyed the impacting body, and would have ejected a ring of rock and dust around the early Earth. Part of this ring then accreted to form the Moon, while other debris from the collision fell back to Earth, or onto the Moon, or was flung out into space by a sort of gravitational sling-shot effect. This theory currently meets all the planetological criteria, both from the point of view of chemistry and of celestial mechanics. To obtain a definitive answer to the enigma of the origin of the Moon, it is essential for scientists to make a return visit. But, thirty years after the Apollo 17 mission, projects for estab-lishing scientific bases on the Moon have still not been financed by the major international space agencies.

Today, in six regions of the Moon, large silvery beetles with blackened feet bear silent witness to landings by spacecraft from Earth. For kilometres all around them, traces of the astronauts' footprints are to be seen in the ashen soil. The Moon will preserve them for a long time. With no atmosphere, the surface has no wind, no rain, no clouds, and no seasons. Days and nights follow one another at a very slow pace, under an unchanging sky. The days are perpetually sunny, and all have been identical for tens of millions of centuries. Because the Moon has not the slightest erosion to efface traces of its history, except for the unceasing rain of micrometeorites and cosmic rays, these footprints will remain in the lunar soil for millions of years, perhaps long after the species that left them has disappeared.

Mars:

a trip to the desert planet

■ RED MARS IS UNDOUBTEDLY THE MOST FASCINATING OF
ALL THE PLANETS. ALTHOUGH SMALLER THAN THE EARTH, IT
CLOSELY RESEMBLES IT. ITS SURFACE, WHERE STREAMS ONCE
RAN, IS COVERED BY FIELDS OF DUNES THAT RESEMBLE
THOSE OF THE SAHARA, AND HIGH VOLCANOES THAT
EVOKE THOSE IN THE HIGH ATACAMA DESERT. MARS
ALSO EXPERIENCES SEASONS, AND THE CYCLIC
MOVEMENT OF GLACIERS. PERIODICALLY IT IS
COVERED IN FOGS AND CLOUDS, AND
OCCASIONALLY STORM-FORCE WINDS
BLOW ACROSS ITS FREEZING DESERTS.

■ IN CHRYSE PLANITIA, WINDS DOMINATE, HAVING REGULARLY BLOWN AT STORM FORCE FOR MILLIONS OF YEARS, SLOWLY SHAPING THE BASALT ROCKS THAT ARE STREWN ACROSS THE MARTIAN DESERT.

The desert is motionless. A stony plain, scattered with a myriad angular rocks, occasionally shaped like tiny pyramids, extends as far as the eye can see. In places the blocks are partially buried in piles of sand. Elsewhere, they are covered by a fine layer of orange dust. They have a tawny shade, verging on brown. They look like basalt – this rock is of volcanic origin. The whole landscape has the same orange tint, and even the sky is salmon-pink. Towards the southwest there is a large bowl with a prominent rim, looking rather like one of the craters in the Auvergne or Arizona. Curls of orange dust raised by the wind obscure the distance. Along the horizon, above the crests of a series of dunes, the sky, first pink, then deep blue, shades into black at the zenith. The gale-force wind – in absolute silence – fails to ruffle the dunes. Nothing moves. As the day comes to an end, the shadows that the rocks cast on the sand lengthen. The temperature has dropped to −78 °C. It is dusk on a fine winter's day, and the Sun is about to set behind the horizon on the Utopia plain on the planet Mars.

It is night. The sky glitters with stars. The Milky Way, gleaming like an immense arch from one horizon to the other,

slowly swings towards the west. Deneb, Vega, and Altair blaze with a steady light; Antares seems too bright and too red. Here, the stars do not seem to have that friendly rapport that they have always had with human beings: no twinkling, no secret significance. Here there are too many stars in the sky for anyone to discern specific constellations: the night is devoid of any images and signs.

Every $7^h 40^m$ and $30^h 20^m$, the satellites of Mars race across the sky like two brilliant meteors. It is a puzzling sight: Phobos, in fact, completes its orbit much faster than one martian day, and is unique in the Solar System. This, the larger of the martian satellites, appears to move 'backwards' in the night sky: it rises in the west and sets in the east. Its speed, close to a degree a minute, is such that the motion is visible to the naked eye. Each day, Phobos crosses the martian sky three times, passing Deimos which, much slower, crosses the heavens from east to west. The two tiny martian satellites are quite incapable of illuminating the martian night as the Moon does for the Earth. The surrounding landscape is invisible, a mere shadow play along the horizon. With the sky, its thousands of stars, and the crests of the dunes behind which the Milky Way

■ This extraordinary view of the Argyre basin, excavated 4 billion years ago by the impact of an asteroid, clearly shows the high-level clouds in the extremely thin Martian atmosphere above the southern hemisphere's high, cratered plateaux. Aeolian erosion has not destroyed the Argyre basin, nor the smaller craters around it. Elsewhere, volcanism, the slow movement of glaciers or dunes have caused traces of Martian history to disappear.

is setting, it would be easy to believe oneself lost in some remote corner of the Sahara. But the stars do not twinkle, nor do they cluster together in the form of familiar animals or mythical beasts and, when a faint violet glow indicates the approach of dawn, the temperature is around −130 °C.

With a diameter of 6794 km, Mars is almost half the size of the Earth. Like the Earth, it rotates on its axis in one day (24^{h}40^m, to be precise). Like the Earth, its rotation axis is inclined at 24°, which results in four seasons, as on Earth. Again like the Earth, Mars has poles that are covered in ice caps, which retreat in summer and grow in winter. Above all, however, Mars is a desert planet: its surface, which is equal in area to that of all the terrestrial continents, is an immense expanse of sand and rock, almost completely unchanging, and covered by a tenuous atmosphere. The air on Mars is, in fact, between one hundredth to one thousandth of the density of that on Earth, and consists almost exclusively of carbon dioxide. It is both tenuous and unbreathable.

In the early morning, after a particularly cold night, the Utopia plain begins to be illuminated and warmed by the Sun. It reveals a strange sight: it has snowed on Mars. More or less all over the surface deposits of frost cover the rocks, and have given a white hue to the reddish sands. It is actually a mixture of water ice and carbon-dioxide ice. Because of the extremely tenuous nature of the martian atmosphere, the way the gases have condensed may almost be likened to part of the air having solidified and then, during the course of the night, having literally fallen onto the surface of the planet!

It is 6 o'clock, and the Sun is slowly heating the martian desert. At the zenith, a few stars are still visible. The temperature slowly rises above −120 °C, and the carbon-dioxide ice vaporizes and returns to the atmosphere. A light mist forms above the martian landscape.

The only things that change on Mars nowadays are the atmospheric conditions. Despite its thinness, the martian atmosphere is capable – as on Earth – of condensing into morning mists, high-altitude cirrus, or orographic clouds. Storms,

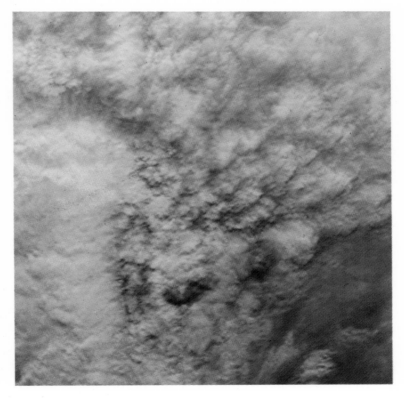

■ WHEN THEY DO OCCUR, THE GREAT MARTIAN DUST-STORMS BLANKET EVERYTHING. THE WIND, BLOWING SOMETIMES AT NEARLY 200 KM/H, RAISES THE FINEST PARTICLES OF MARTIAN DUST AND DEPOSITS THEM HUNDREDS OF KILOMETRES AWAY. NEAR THE NORTH POLE, THE DESERT HAS BEEN COVERED IN FIELDS OF DUNES THAT ARE LARGER THAN THE SAHARA. OVER THE COURSE OF TIME, THIS AEOLIAN EROSION CHANGES THE FACE OF THE PLANET.

and tropical cyclones that so closely resemble terrestrial hurricanes that they might be taken for them, frequently occur on the red planet. Paradoxically, the intense activity in the martian atmosphere is actually caused by the low density of the air. In the middle of the summer, during the course of a single day between early morning and the middle of the afternoon, the temperature can rise from −100 °C to 0 °C! Thermal differences of this magnitude easily disturb the atmosphere's equilibrium and may give rise to genuine cyclones.

The eccentricity of Mars' orbit means that its distance from the Sun varies considerably. In one year (i.e., in 688 terrestrial days), that distance changes from 207 million kilometres to 249 million kilometres. This results in major seasonal effects on the red planet. The most spectacular of these are undoubtedly the changes in the polar caps, which, using even the smallest amateur telescope from Earth, may be seen to vary over the course of the year. During the winter, which lasts on Mars nearly six Earth months, the southern polar cap extends over about 10 million square kilometres. In spring, and then during the southern summer, is slowly melts (or rather, sublimes) and releases a large quantity of carbon dioxide into the atmosphere. By the end of the summer, the polar cap has almost disappeared.

Every year, at the beginning of the martian summer, the heating induced by the position of the Sun in the sky disturbs the martian atmosphere, and strong winds arise nearly everywhere across the red planet. Towards the summer solstice, the increasingly violent winds pick up the finest of the particles that evenly cover the ground. In just a few weeks, this great dust storm envelops the whole planet and covers it in a dense dust haze; a haze that is so dense that, from orbit, the surface of Mars is invisible. In 1976, two of NASA's Viking spaceprobes landed on Mars, on the Chryse and Utopia plains. For several years they photographed the landscape, recorded the temperature and pressure, and observed climatic variations. It was thus that Viking 2 found itself, right in the middle of the martian summer, in a real storm. It was a strange ex-

perience: the probe, resting on its three legs, did not move, did not even vibrate. In almost absolute silence, the wind blew with gusts of 150, 180, and even 200 km/h. Yet nothing moved. The dunes remained static, the sand was not blasted away. On Mars, the atmosphere is so tenuous that the wind doesn't affect anything. Only microscopic, invisible particles are transported. The camera on Viking 2 was, however, able to record the effects of the event: the Sun, hidden by a dense haze, no longer cast any shadows on the surface, and the horizon was slightly veiled. And that was it: the great martian storm. The low rate of aeolian erosion enables planetologists to read the history of the red planet like an open book. On Earth, the extraordinary changeability of the crust, which involves continental drift, rapid erosion through the combined action of water, wind and (above all) life, erases any traces too quickly, and complicates the quest for our origins. On Mars there is no plate tectonics: the red planet is just a single vast continent with a stable crust. And although erosion has drastically altered the landscape in places, in other regions geographers and geologists discover that the planet remains as it was 3 or 4 billion years ago. The southern hemisphere displays the oldest and most uniform landscape on Mars: over an area of more than 100 million square kilometres impact craters of all sizes pepper the surface of the planet. The smallest measure a few tens of metres across, or even less, and the largest are some hundreds of kilometres in diameter. This incredible bombard-

■ THE MOUTHS OF ANCIENT
LAVA FLOWS ON THE AMAZONIS
PLAIN. A YOUNG IMPACT CRATER
IS CLEARLY VISIBLE AT TOP LEFT.
IT IS SURROUNDED BY TYPICAL
LOBED EJECTA. AS THE
METEORITE HIT THE SURFACE,
THE SUBTERRANEAN ICE
MELTED ABRUPTLY AND
CREATED AN ENORMOUS
WAVE OF MUD, WHICH
SOLIDIFIED AFTER
TRAVELLING SEVERAL
KILOMETRES.

ment is evidence of the origin of the Solar System, when billions of asteroids criss-crossed space in all directions and, from time to time, crashed into a planet. Some of the scars on the red planet will never be erased, such as the Hellas basin, which is more than 1400 km in diameter and 6000 metres deep. Some regions of Mars, such as Noachis and Eridania, have remained in practically the same state for 4 billion years, which is an incredibly long period. On our own planet, vestiges of this distant past, lying at the very dawn of the Archaean era (known as the Noachian on Mars), are confined to a few rocks found in deep bores in Greenland or the Canadian Shield.

Despite the extraordinary age of its oldest regions,

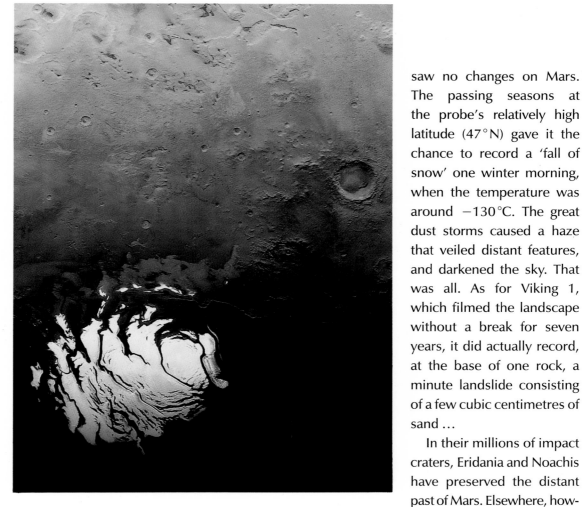

■ LIKE THE EARTH, THE PLANET MARS HAS TWO POLAR CAPS. HERE, AT THE HEIGHT OF THE SOUTHERN SUMMER, THE SOUTHERN CAP FINISHES ITS RECESSION. UNLIKE THE NORTHERN CAP, WHICH PRIMARILY CONSISTS OF WATER ICE, THE SOUTHERN POLAR CAP MAINLY CONSISTS OF FROZEN CARBON DIOXIDE. ITS SPIRAL SHAPE IS WITNESS TO THE VIOLENCE OF THE MARTIAN WINDS THAT FLOW ROUND THE POLE.

Mars is not the Moon or Mercury, and the surface of the planet has continued to evolve since the Noachian era. Over a long period, storms have eroded the surface, blurring the impact craters, raising dust from the high plains and depositing it at the bottom of winding valleys, creating dune fields in various areas, and slowly clearing certain plateaux, such as Syrtis Major, only to cover the Hellas basin in a fine coat of rust. In the long term, the landscape has changed. The great plains of Chryse and Utopia, where the two Viking Landers touched down, show traces of this aeolian erosion everywhere, from the heaps of sand to the rocks that have become facetted into tiny pyramids by the silent winds that have been raging since the beginning of time. But the process is a long one, an extremely long one. Viking 2, during the four years that its mission lasted,

saw no changes on Mars. The passing seasons at the probe's relatively high latitude (47°N) gave it the chance to record a 'fall of snow' one winter morning, when the temperature was around −130°C. The great dust storms caused a haze that veiled distant features, and darkened the sky. That was all. As for Viking 1, which filmed the landscape without a break for seven years, it did actually record, at the base of one rock, a minute landslide consisting of a few cubic centimetres of sand …

In their millions of impact craters, Eridania and Noachis have preserved the distant past of Mars. Elsewhere, however, the planet has been subject to intense activity, unique in the whole Solar System. After the phase of unrelenting bombardment of the Noachian era came the Hesperian era, during which the planet slowly created titanic geological structures that formed a sort of monumental architecture. Mars had plenty of time. Here, there are no crustal movements, no continental drift, and practically no erosion; the mountains and valleys would persist for ever. In addition, the martian gravity is half that of Earth's. Here again, the mountains – or rather the volcanoes, because there are no equivalents to the Andes, the Alps or the Himalayas on Mars – were able to benefit.

It is early morning on Olympus Mons. It is a dreary plain of grey basalt, ancient flows of ropy lava, powdered with orange sand, and harshly lit by a Sun, which although hardly above the horizon, seems to be unusually brilliant. The cold is terrible, −120°C, and the sky,

■ DAYBREAK ON MARS. IN THE FIRST LIGHT OF DAWN, A THIN CARBON-DIOXIDE FOG SWATHES THE HORIZON OF UTOPIA PLANITIA. HERE, AT THE BEGINNING OF THE MARTIAN WINTER, THE TEMPERATURE REACHES −118°C. WHEN THE SUN RISES, THIS FOG WILL DISPERSE, THE SKY WILL CLEAR AND IT WILL BE A FINE DAY, ALTHOUGH FREEZING COLD.

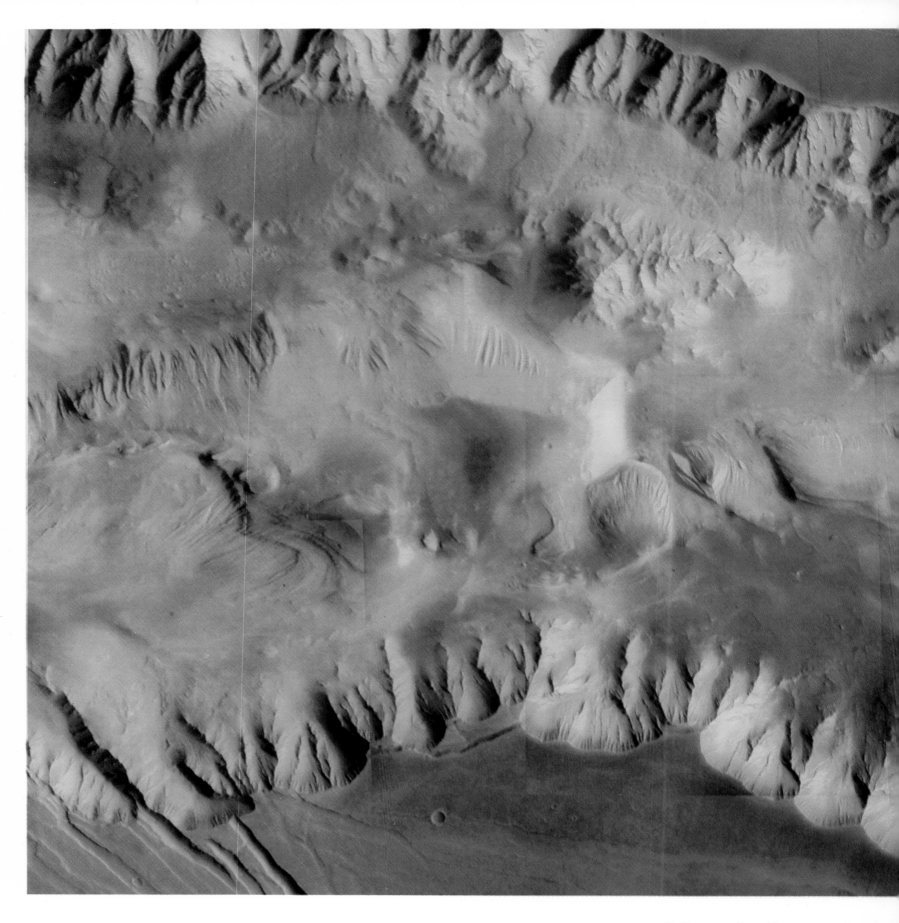

■ CLOSE-UP OF VALLES MARINERIS: IN THE COPRATES
CHASMA REGION, THIS GIGANTIC CANYON, 200 KM
WIDE, IS FLANKED BY CLIFFS 7000 M HIGH. FROM
TIME TO TIME THE SIDES OF THIS GREAT MARTIAN
CANYON COLLAPSE IN CATASTROPHIC LANDSLIDES,

orange-coloured along the horizon, changes abnormally
quickly to dark blue. The rocky plain slopes slightly towards
the north, at an angle of 5°, but does not show any particular
form of relief. To the south, our view stretches for a staggering
distance, fading away beyond the line of coloured clouds that
hang silently along the horizon. To the north, nothing but lava
that froze several thousands of millennia ago, that still seems
to want to drain away from the vast blocks of rock that stand,
silent, immobile, and watchful over the scene. High above is
an indigo sky in which a few stars glitter. From the west,
moving contrary to the apparent motion of the stars, Phobos is
about to slowly cross the sky.
As bright as Venus when seen
from Earth, its elongated shape is perfectly detectable with the
naked eye. (As seen from the martian surface, Phobos
measures almost 15 minutes of arc across, about half that of
the Moon seen from Earth.) It will set in the east about four
hours later.

Is this really a perfectly ordinary martian landscape? The
intense cold, the sky, with its crystalline clarity, in which stars
glisten – in broad daylight – does leave a whiff of strangeness.
This gentle slope is in fact part of the flanks of an enormous

90

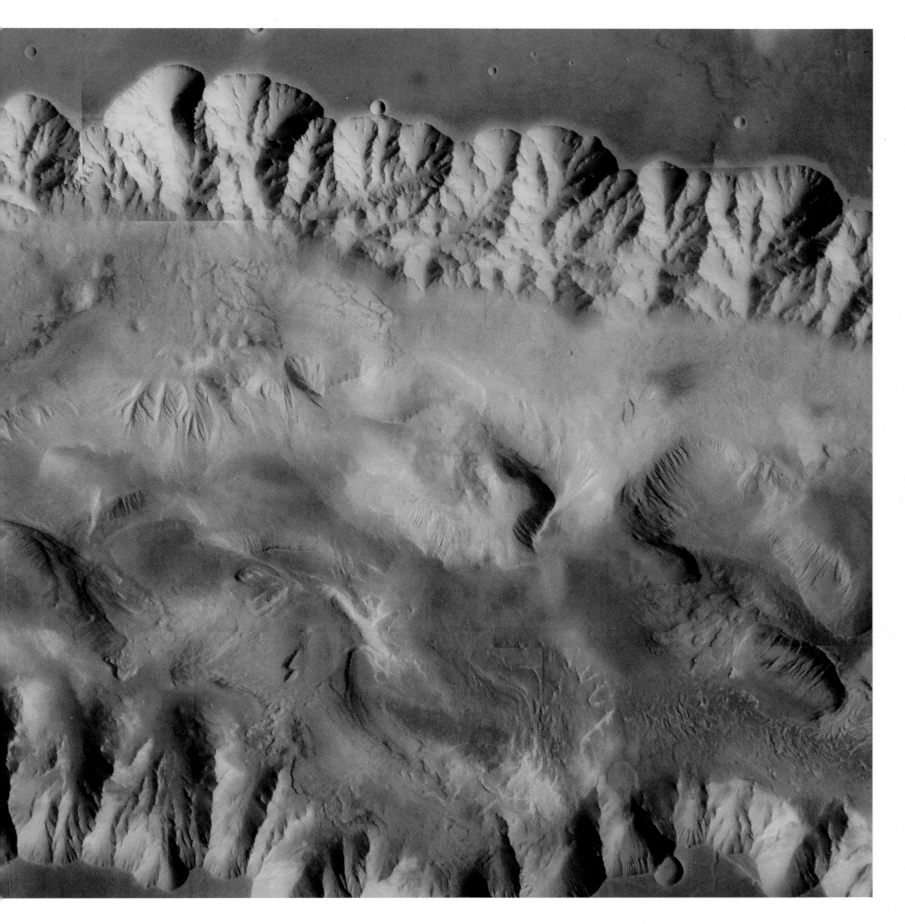

WHICH RELEASE THE WHOLE SIDES OF MOUNTAINS AND WHICH DO NOT STOP UNTIL THEY ARE TENS OF KILOMETRES AWAY AND MUCH LOWER, LEAVING BEHIND IMMENSE DETACHMENT SCARS.

volcano, which is literally the largest mountain in the known universe. Olympus Mons has a height of 27000 m, and is about 1000 km in diameter at its base! At the edge of the summit caldera, looking over the crater, the view is stupefying: the almost vertical walls, 3000 m high, that encircle it stretch away in a vast parabolic arc and are lost in the distance, meeting 80 km farther on, way beyond the horizon. There is no geological edifice anywhere in the Solar System that is as gigantic as Olympus Mons. It, like the three other giant volcanoes that accompany it – Pavonis Mons, Arsia

Mons, and Ascraeus Mons – owes its existence to the weak gravity on Mars and, above all, to the lack of any overall motion of the crust. The lava, rising from the depths in the very distant past, pierced the crust at what geologists call a 'hot spot'. Then, over a period of tens or hundreds of millions of years (or even billions of years) it spread out at the surface, slowly building the volcanic edifice.

The great martian volcanoes do not lie far from the planet's equator, in the region of the Tharsis Ridge. This is an immense dome, some 6000 km in diameter, where the martian crust has been raised by about 7000 m. Did this Tharsis bulge open up

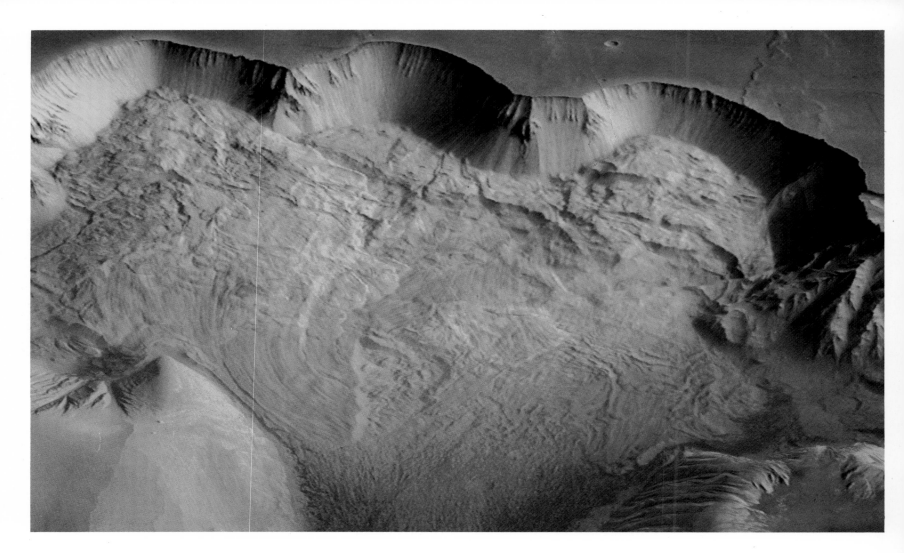

the surface of the planet? Or did Mars attempt to initiate plate tectonics in this region?

The Tharsis Ridge is crossed by an immense chasm, nearly 5000 km long, formed by the network of canyons of Valles Marineris. It is a landscape worthy of giants: when the Sun rises at one end, around Coprates Chasma in the east, Tithonium Chasma in the west is still plunged in the deepest night. The origin of the Valles Marineris system remains unknown to planetologists. It is as long as the Great Rift Valley in Africa, or the Mid-Atlantic Rift, and 250 km wide. On both sides of the trench there are cliffs between 3000 and 7000 m high. The size and imposing majesty of this natural wonder have no equivalent on Earth, nor on any known planet. Ever since its recent origin – relatively speaking – perhaps only 1 or 2 billion years ago, the Valles Marineris system has continued to enlarge. Patient erosion by the wind (the unceasing breath of the planet) that constantly blows through its canyons has caused avalanches from time to time. Imperceptibly, the Valles Marineris cliffs continue to recede. From the edge of Coprates Chasma, the deep-

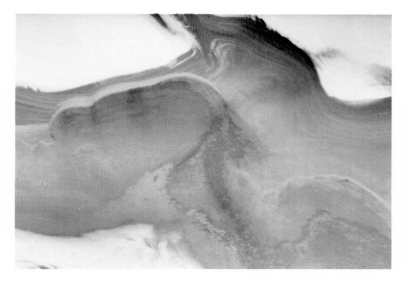

■ At Mars' North Pole, the ice cap advances and retreats with the seasons. This ice movement, together with the repeated dust storms, which are particularly violent in these dune-covered areas, deposit layers of sediment every year. The layers of ice and dust are clearly seen here.

est and most impressive of all the long series of collapse features that mark out Valles Marineris, it is difficult to make out the narrow orange line of the opposing cliff, half hidden below the horizon of this small planet. By contrast, flying over Valles Marineris reveals a fantastic sight: all along the sides of the giant martian canyon landslides have removed whole sections of the cliffs, leaving behind immense detachment scars. In Coprates Chasma, the largest of these avalanches caused a slice of mountain 60 km long to collapse as a single unit. It is difficult to imagine such an apocalyptic scene, which took place without a single witness. With a dull rumbling sound, thousands of billions of tonnes of martian rocks hurtled down the side of the canyon at 200 km/h, before coming to rest, half-an-hour later, 7000 m lower and 100 km farther on.

Imagine the scene much farther north, on a fine summer's day, at the foot of a great barchan dune, somewhere in Vastitas Borealis. The Sun, at this very high latitude (75° N) illuminates the landscape throughout most of the day, but culminates very low above the horizon. The light has an

THE NORTHERN HEMISPHERE, IN THE CERBERUS, ELYSIUM AND AMAZONIS REGION, IS ALMOST COMPLETELY DEVOID OF CRATERS. IT IS ONE OF THE YOUNGEST AREAS OF THE RED PLANET. RIGHT IN THE UPPER LEFT CORNER OF THE IMAGE IS THE SMALL CRATER MIE, WITH ITS DARK CENTRAL PEAK. VIKING 2 LANDED IN THE UTOPIA PLAIN, ON THE LEFT-HAND EDGE OF THIS CRATER.

exceptional quality; the great dunes, lit through the orange-coloured haze of the lower atmosphere, glow with the red and tawny colours of the martian dust. There are thousands of these barchans trailing their golden crescents across the ergs of the Great Northern Desert, advancing infinitely slowly across the surface, following the rhythm of the seasons on the red planet. On Mars, the equivalent of the Sahara occurs around the North Pole. These ergs, with their barchan and transverse dunes, have completely wiped away the ancient history of the planet. Here there are no craters, no volcanoes, nothing but lines of dunes like the waves of some frozen ocean. With its northern deserts where, in winter, the temperature drops below −140 °C, yet again Mars shows one of the extremes that have no equal elsewhere in the Solar System: the innumerable dunes in these great ergs cover an area of nearly one million square kilometres.

Mars has not always been this desert of dust and sand, where the only obvious activity is that shown by the clouds of carbon-dioxide ice rising up the flanks of Olympus Mons, or the slow abrasion of the northern plains beneath the silently advancing dunes, driven by the great equinoctial storms. If we return to temperate latitudes, towards the Utopia and Chryse plains, where the winds have not effaced martian history and, above all, to an area where the latter does not simply consist of the ancient remnants of thousands of impact craters, we find a strangely familiar landscape: that of Lunae Planum. This piece of red desert could be in the Hoggar or the Takla Makan and leaves an impression of *déjà vu*. Is it that steep-sided valley, striped with sedimentary layers, which fade into the pink-tinted distance? Is it those deposits of alluvium, which have filled the canyon where we are; or these traces of erosion, which have obviously followed a sinuous course, here avoiding a large lump of rock, and there hollowing out a bend at the foot of an orange cliff? Everything here, in Maja Vallis, which cuts across Lunae Planum for more than 1000 km, suggests the passage of flowing water. This martian landscape inevitably reminds one of a dried-up wadi somewhere in the Sahara. As on Earth, water must have flowed here.

Yet there is no water, today, on the surface of Mars. At least, there is no liquid water, because the incredibly dry and rarefied atmosphere does not permit it to exist. On Earth, water exists in its three phases: solid, liquid, and gaseous. The famous 'triple point of water' at which water exists in all three forms lies at 0 °C, at a pressure of 1000 milli-bars. But the pressure of the martian atmosphere is less than 10 millibars. If one were to attempt to pour a glass of water onto the surface of Mars, the water would evaporate before it touched the ground. An ice cube left to melt in the Sun at the equator, in high summer when the temperature sometimes exceeds 0 °C, would sublime without even dampening the parched, sterile sand beneath it. Mars is a dehydrated planet, where, paradoxically, the traces of flowing water are visible more-or-less everywhere, including at the bottom of the canyons and in the wadis, which have remained dry ever since.

At the end of the Noachian era, about 3.8 billion years ago, when cratering had more or less come to an end, there was a violent outburst in volcanic activity. The young planet, slowly cooling, spewed out a dense, searing atmosphere through a thousand volcanic vents. A mild, temperate climate must have been established on the red planet, protected by the thick atmosphere of carbon dioxide. As on Earth where, at the same epoch – with an atmosphere that would be completely unbreatheable for modern-day terrestrials – life appeared in the 'primordial soup' of the oceans, the first drops of rain must have started to fall on the vast expanses of solidified lava. First rain, then trickles of water finding their way between the basalt boulders, flowing together into streams, then in torrents carving gullies in the plateaux, in channels red with mud, in rivers as wide as the Nile, and probably eventually ending in vast lakes of placid water. If this Eden lasted long enough, and if the atmosphere contained enough vapour, these could have been true seas. The red planet may have possessed seas in the distant past. Some researchers have even tried to establish where these small oceans may have

OLYMPUS MONS IS THE LARGEST MOUNTAIN CURRENTLY KNOWN IN THE SOLAR SYSTEM. THIS GIGANTIC VOLCANO MEASURES ABOUT 1000 KM IN DIAMETER AT ITS BASE, AND IS 27 KM HIGH. BECAUSE THE CRUST OF MARS IS NOT SUBJECT TO PLATE TECTONICS AS IT IS ON EARTH, THE GREAT VOLCANO WAS ABLE TO GROW OVER BILLIONS OF YEARS.

been situated. The Argyre and Hellas basins lie well below the average level of the martian surface. It is there that true sea level may have been located, if water did ever persist on the surface of Mars. In the northern hemisphere, an enormous region in Vastitas Borealis and Acidalia Planitia could also have been completely submerged. This hypothetical vast expanse of water has even been given a name by planetologists: Oceanus Borealis. If conditions did actually occur as suggested by these researchers, these martian seas, on which pink and gold ice floes may have slowly drifted during the winter, may possibly have been several hundred metres in depth.

But where has the martian water gone? When did it stop excavating the martian canyons? To try to answer this question, twenty-one years after the two Viking probes, we have sent a new ambassador to the red planet: Mars Pathfinder. The small spaceprobe, after a speedy trip – taking just ten months in transit – landed on Mars on 4 July 1997 at latitude 19°33′N, longitude 33°55′W, at the junction of two collapse-type valleys, Ares Vallis and Tiu Vallis. Every day for more than a month Mars Pathfinder transmitted back to Earth images of the desert, where, 3 billion years earlier water had flowed. As with Viking Landers 1 and 2, which had been the pioneers, Mars Pathfinder endured extreme meteorological conditions, although slightly more clement ones than those that reign on the high plains of Chryse and Utopia. At its site in Ares Vallis, Mars Pathfinder's weather station recorded maximum daytime temperatures of −12 °C to −14 °C, and minimum nocturnal ones of −68 °C to −74 °C. Above all, however, its camera, perched on a mast 1.5 metres tall, was able to record the whole panorama around it. And what a panorama! Everywhere, right round the probe, there was a chaotic jumble of brown rocks, with sharp edges, just like the sites discovered by the Viking probes. Farther away, at about 1200 m, there were two hills with distinct summits, named Twin Peaks, and then, sometimes rendered indistinct by

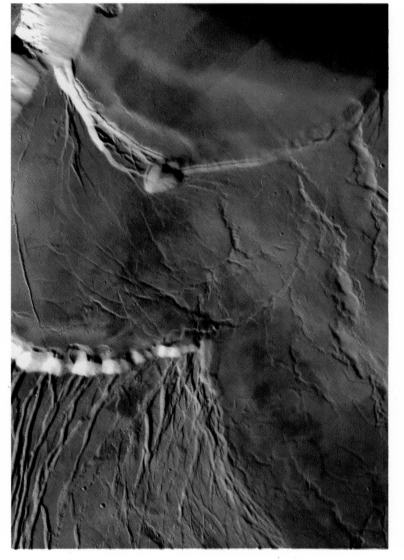

■ THE SUMMIT CRATER OF OLYMPUS MONS, 80 KM ACROSS. THE SUCCESSIVE FLOWS AND ANCIENT LAKES OF LAVA, PRODUCED DURING MAJOR ERUPTIONS, ARE STILL VISIBLE, FRACTURED IN PLACES AS A RESULT OF EARTHQUAKES. THE CRATER'S RAMPARTS ARE MORE THAN 3000 M HIGH. OLYMPUS MONS IS PROBABLY NO LONGER ACTIVE.

the hazes in the lower atmosphere, some indications of distant relief: the rampart surrounding an impact crater, 2.2 km distant; a gentle hill at 21 km; and finally, scarcely discernible on the horizon, a peak several hundred metres high, and 30 km distant.

Ares Vallis, at the mouth of which Mars Pathfinder landed, is the presumed bed of ancient floods, a broad collapse-type valley, some 1500 km long, 25 km wide, and several hundred metres deep. According to the data gathered by the Viking probes, planetologists believe that it experienced one or more episodes of short, catastrophic flooding, some 3 to 3.5 billion years ago. The images taken by the probe do indeed show rounded shapes. Some of the rocks have softer outlines: their angles are not sharp and seem to be worn, perhaps as if they had been transported by water. According to the geologists, there is not the slightest doubt that there are less angular rocks at Ares Vallis than at the two Viking sites. In addition, there are also very flat rocks, which look as though they consist of sediments. The probe's images also show that there are numerous rocks with parallel faces embedded in the soil, all inclined at the same angle, suggesting the presence of a strong current. If Ares Vallis had been eroded by a glacier, rather than by liquid water, these rocks would have been deposited in random positions. The data from Mars Pathfinder therefore seem to confirm the theory that there were floods of liquid water in Mars' past, rather than that there was erosion by ice. But it was also confirmed that the valleys on Mars did not undergo prolonged flooding: there is not a single pebble among all the rocks at Ares Vallis.

Following Mars Pathfinder, which survived the red planet's severe climate for nearly three months, two new probes, Mars Climate Orbiter and Mars Polar Lander, were due to continue martian exploration in 1999, the first from orbit, and the second from the surface. Unfortunately both of these NASA probes were 'lost with all hands' in the vicinity of the red

planet. The first failed to be inserted into orbit, because of a human error in spatial navigation, the second crashed onto the surface. This was a great shame, because the landscape that Mars Polar Lander was due to examine was radically different. The probe was, in fact, scheduled to land not far from the South Pole, at latitude $-76°$, which corresponds, on Earth, to the middle of Antarctica. It is a region where sand and ice fight for supremacy, depending on the martian season. These two failures added to the loss of two other probes, Mars 1996 and Mars Observer, which occurred earlier in the 1990s.

Early in 2001, there was just one survivor in martian orbit. Mars Global Surveyor (MGS) which, since 1999, has been regularly photographing the planet from an altitude of 450 km and has returned more than 100 000 images. Like Mars Pathfinder, as yet Mars Global Surveyor has not been able to support the hypothesis of a martian past that was extremely rich in

■ PART OF THE VIKING 2 LANDER PHOTOGRAPHED WITH ONE OF ITS CAMERAS. THE BASE OF THE ANTENNA THAT TRANSMITTED THE SIGNALS BACK TO EARTH IS VISIBLE IN THE FOREGROUND, SILHOUETTED AGAINST THE SKY.

water. In reality, some of the measurements, particularly those recorded by the American probe in 2000, are contradictory. On the one hand, they weaken the poetic, but fragile theory for Oceanus Borealis, the ancient martian ocean. According to the data, the 'shorelines' that Viking had apparently discovered, do not actually exist.

Observed at very high resolution by MGS, they literally vanish into the desert sands. On the other hand, the data recorded by the probe's laser altimeter rather tend to confirm the oceanic theory. The altimeter, which is capable of detecting difference in altitude of just a few tens of metres, may be able to reveal the presumed coastlines of the ancient martian oceans. And, in fact, MGS has discovered topographical levels that do actually resemble ancient tide-lines around basins that lie not far from the North Pole. According to planetolo-

■ THE SURFACE OF ANOTHER WORLD: UTOPIA PLANITIA. THIS NORTHERN PLAIN WAS DEVASTATED BY A METEORITIC IMPACT A FEW HUNDRED MILLION YEARS AGO. THE DUSTY SOIL HAS BEEN EXCAVATED BY THE VIKING 2 LANDER'S REMOTE-CONTROL

gists' calculations, the volume of water would have reached at least fifteen million km³, which represents, at the low-water level, the volume of the Atlantic and, during the red planet's ancient warm periods, more than ten times as much. But where has the martian water gone? When did it cease to erode the canyons of Valles Marineris?

These questions were destined to be answered by the lost spaceprobes. Currently, however, the programme of martian exploration has been disrupted by the succession of failures that occurred between 1996 and 2000. There is, in fact, a favourable 'launch window' for Mars that corresponds to the few weeks during which the Earth is approaching Mars, and which occurs every twenty-six months. Outside these short periods, the two planets are too far apart for anything to travel between the two in a reasonable period of time.

■ THE SAME UTOPIA PLANITIA LANDSCAPE IN THE MIDDLE OF WINTER. A FINE LAYER OF CARBON-DIOXIDE SNOW HAS BEEN DEPOSITED ON THE SOIL AND MAY BE SEEN MORE OR LESS EVERYWHERE ON THIS PHOTOGRAPH. THE TEMPERATURE IS ABOUT −120 °C.

The next launch window in 2001 was used by NASA to launch Mars Odyssey 2001, a probe designed to establish a geological map, to search for possible traces of ice in the subsoil, and to serve as a relay satellite for forthcoming martian missions.

NASA is now preparing the Mars Rover 2003 mission, consisting of two rovers, capable of travelling up to 100 m per day, for a period of at least 3 months. It is probable that, like Mars Pathfinder, Mars Rover 2003 will visit an ancient river bed.

The search for present-day water, in the form of ice this time, is the objective of the first European mission to the red planet: Mars Express, which will leave the Earth on board a Semiorka rocket in August 2003. This very ambitious mission consists of two separate spaceprobes. The first is an observation satellite, which will orbit Mars for more than a year, systematically looking at and photo-

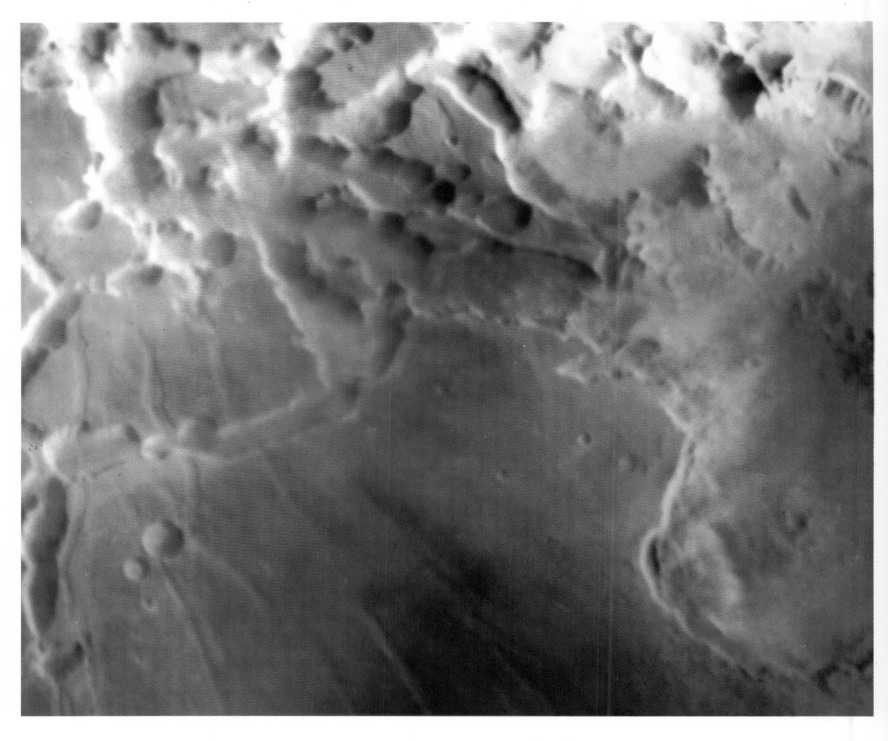

■ TOWARDS THE WEST, THE IMMENSE NETWORK OF CANYONS OF VALLES MARINERIS RUNS INTO NOCTIS LABYRINTHUS. THIS IS A TANGLE OF ENCLOSED VALLEYS THAT CUT INTO THE THARSIS PLATEAU. ON THIS IMAGE, THE DEEP VALLEYS HAVE BEEN INVADED BY EARLY-MORNING MISTS.

graphing different areas, analyzing the mineralogy and geology of the surface, and searching for traces of ice in the subsoil by means of radar. The second is Beagle 2, a lander that will set down on a plain in the northern hemisphere that was once flooded by catastrophic flows of mud. There, a colour camera will photograph the whole of the landscape. Then Beagle 2 will deploy a 'mole', which will obtain samples from below the surface and deposit them into a mini-laboratory, to search for possible fossil life-forms or bacteria. In fact, the scientists have no illusions about the chances of finding life on the red planet.

Mars is not the Earth. One tenth of the latter's mass, and more distant from the Sun, the red planet has seen its atmospheric system slowly become transformed. Part of the atmosphere has escaped into space; the water vapour has frozen into ice, has become trapped at depth with the last rains, and has slowly migrated towards the poles as immense glaciers which have scraped out the valleys in the north over millennia, and which have then, in their turn, sublimed and disappeared. There is still water on Mars, but it is hidden as ice mixed with rock, like the Siberian permafrost, hundreds of metres below the surface. It freezes into thin cirrus clouds that lap the flanks of the great volcanoes and which slowly drift over Utopia Planitia. Or, permanently frozen, it forms the ice-fields and glaciers of the two martian polar caps. Since the Noachian era, the red planet's wadis have channelled nothing more than rock dust. And yet – from time to time, as the result of a geological upheaval or a new climatic cycle, the faint echo of water running and bubbling before it freezes and then evaporates may be

■ MARS SEEN FROM EARTH ORBIT BY THE HUBBLE SPACE TELESCOPE. AS IN THE PHOTOGRAPH ON PAGE 85, HERE AGAIN THE STRIKING SHAPE OF SYRTIS MAJOR IS VISIBLE.

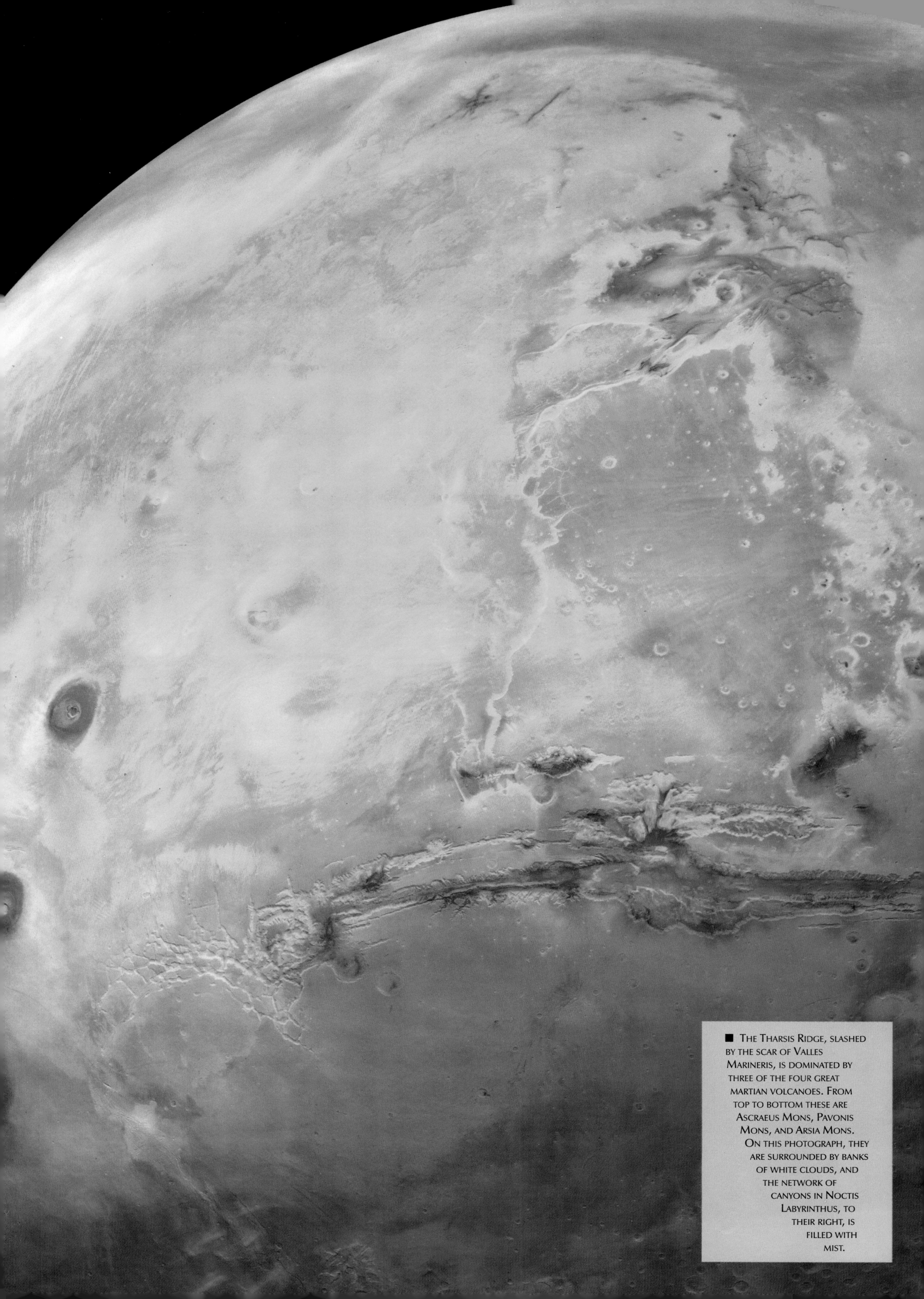

■ THE THARSIS RIDGE, SLASHED
BY THE SCAR OF VALLES
MARINERIS, IS DOMINATED BY
THREE OF THE FOUR GREAT
MARTIAN VOLCANOES. FROM
TOP TO BOTTOM THESE ARE
ASCRAEUS MONS, PAVONIS
MONS, AND ARSIA MONS.
ON THIS PHOTOGRAPH, THEY
ARE SURROUNDED BY BANKS
OF WHITE CLOUDS, AND
THE NETWORK OF
CANYONS IN NOCTIS
LABYRINTHUS, TO
THEIR RIGHT, IS
FILLED WITH
MIST.

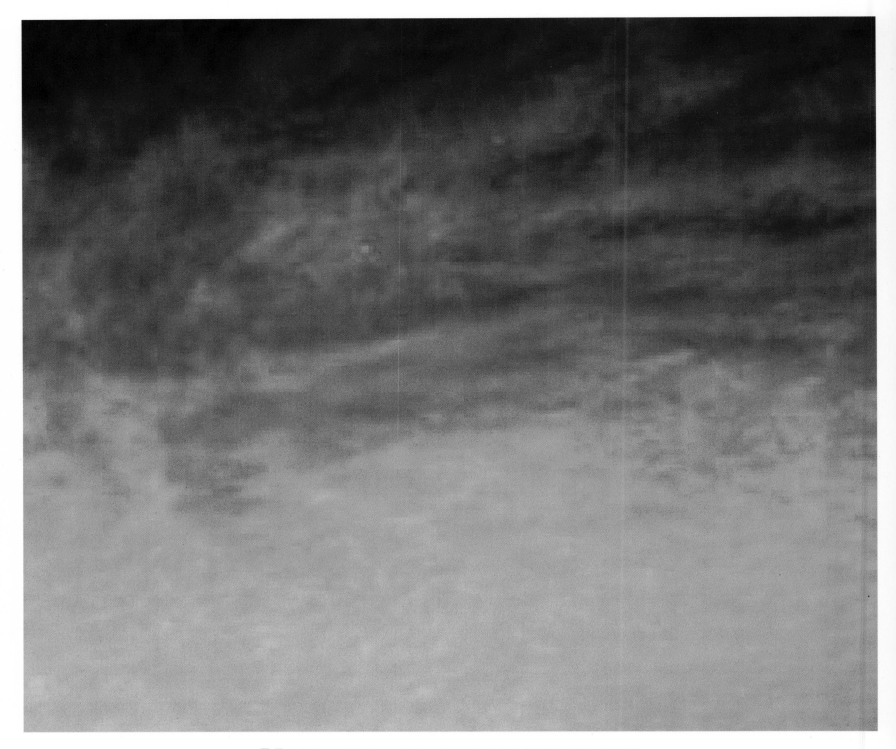

reflected from the cliffs of the giant canyons. In places on Mars, especially in the younger northern hemisphere, planetologists have discovered traces of relatively recent flows. These fossil rivers, very wide and very short, arise in the volcanic regions, in the midst of chaotic heaps of mountainous blocks, which suggest gigantic landslides. What has happened here?

Let us step back several million years, to the era when, on Earth, *Homo erectus* began to stand upright, and was about to become *Homo sapiens*.

It is early afternoon in Ravi Vallis, at the beginning of summer, towards the end of the Amazonian era. Phobos and Deimos are visible through the orange-tinted, clear air, peacefully crossing the sky. Since the morning, the temperature has risen above 5 °C, and the pressure is 100 millibars. Mars, at the closest point to the Sun in its overall cosmic cycle, has almost regained the idyllic conditions that occurred at the end of the Noachian era. We are in a valley, 20 km wide, enclosed by great walls, 3000 metres high. Everywhere there is the same landscape of brown boulders, banks of orange-tinted sand, and millions of basalt rocks buried in the pale dust, left there following the impact of an asteroid, 4 billion years earlier. Water glistens everywhere: it seeps out of the ground, forming minute puddles which slowly evaporate, it darkens the pale sand, and it freezes in the shadow of the volcanic rocks. Suddenly, something abnormal happens. There is a vague rumble, a dull vibration that causes the ground to shake, momentarily troubling the landscape. That's all. Then there is nothing but that tense silence that precedes disaster. Tiny threads of water continue to seep between the stones. Then, as the ground starts to shake, and an indescribable rumbling fills the air, a towering red wave, 80 m high, appears on the horizon. Mud, water, ice, and rocks all mixed together, overwhelm everything, sweep everything away, carrying it downstream, gouging out Ravi Vallis, finally turning into a sort of monstrous cloud, loosing its impetus and disappearing into the cracks in the rock, swallowed up by the sand, evaporating, and finally freezing after having travelled only 120 km across the martian desert. A disaster that has no equivalent on Earth, and the

scars from which Mars has retained, probably for ever.

On Mars that day, a 10 000 km² plateau suddenly collapsed. Swollen with melted ice – doubtless because of subterranean heating – it abruptly released an enormous torrent, amounting to a cubic kilometre of water per second (more than one thousand times the volume of the Amazon in flood). Then, just as swiftly as the gigantic wave had appeared, the water disappeared, leaving behind for the cameras of the Viking probes the strange trace of an enormous wave that had dried up and left no distinct signs of how far it had reached.

Elsewhere, later, an ancient subterranean lake perhaps collapsed, suddenly giving way to the weight of the ground, erupting as a geyser at the surface, feeding for a few minutes a wadi that had been dry for 100 million years, before leaving the area, for another 200 or 300 million years, to the dunes of sand and dust. So, at very rare intervals, either as a result of a closer approach to the Sun, or because of magma rising from the planet's interior, water may flow on Mars. Even today, liquid water may perhaps exist on the desert planet, somewhere near the equator, in Ares Vallis, in Sinus Meridiani, or even in the Hellas basin.

These regions are actually the lowest on the planet. In the absence of any ocean, planetologists have chosen an arbitrary reference level, which corresponds to an atmospheric pressure of 6.1 millibars, and which is moreover also the pressure that, at a temperature of 0 °C, corresponds with the triple point of water. In Sinus Meridiani, which lies nearly 2000 metres below the reference level, the atmospheric pressure should attain about 10 millibars (as against 2 millibars on the highest volcanoes). When the temperature in summer at the martian surface rises above 0 °C, the underground ice – especially if it is salty – should melt easily and may perhaps remain stable in the liquid form. Can we not imagine that beneath the planet's surface there are caves and caverns that shelter subterranean lakes where fresh water, heated by some thermal source, and protected from the desiccated surface by a coating of ice or by permafrost, awaiting better days to burst forth and flood out into the sunlight?

And the burning question that is always in the background

■ ONE OF THE MOST BEAUTIFUL PANORAMAS OBTAINED BY THE MARS PATHFINDER PROBE IN 1997. ON THIS IMAGE OF THE LANDSCAPE AT ARES VALLIS, TWO HILLS (KNOWN AS TWIN

of our minds returns when faced with these far from extra-terrestrial landscapes – which seem so familiar that every time we turn the corner of a dune we expect to see a camel munching the branch of an acacia. And that is: What about life? Why, under initial conditions that were so close to those on our own planet, should it not have appeared on Mars as well? Mars, of all the planets in the Solar System, is probably the one that, with Earth, is the most welcoming. And yet the Viking and Mars Pathfinder probes disappointed us in this search for extraterrestrial life. From martian orbit they saw nothing that resembled traces of life: no tracks left by migrating animals, no cyclopean architecture, and no ancient cities buried in the sand. On the ground, in the northern hemisphere's great plains, their cameras did not spot the slightest furtive movement, the tiniest insect, or the smallest blade of grass. Their chemical and biological detectors, which provided complex and ambiguous results, did not leave much hope of finding even a bacterium or a virus on Mars. But if there is little doubt that there is no life on Mars at present, on the other hand no one can say that it never appeared and evolved in some tropical oasis, or on the temperate beaches of Oceanus Borealis. In that case, the next martian space missions

■ SUNSET OVER ARES VALLIS AFTER A FINE SUMMER'S DAY. THE SILHOUETTE OF TWIN PEAKS IS FADING ON THE HORIZON. SOON THE EVENING WIND WILL RISE AND THE TEMPERATURE WILL DROP TO −70 °C.

PEAKS), POSSIBLY ERODED BY ANCIENT CATA-
STROPHIC FLOWS, APPEAR ON THE HORIZON.
THEY ARE ABOUT THIRTY METRES HIGH AND LIE
LESS THAN 1 KM FROM THE PROBE.

should set out to search in the beds of dried-up rivers, and in the talus heaps of the great martian canyons, not for tiny living animals, but for fossils. Unless...

Unless a miracle has happened on the red planet as well, and life has been able, somewhere, to adapt to the freezing, desiccated winter that gradually came to rule Mars?

After all, on Earth, life has adapted to conditions that are perhaps even more exotic: there are, for example, blind fish, tube worms, and giant clams that live under incredible pressures and temperatures around hydrothermal vents in the oceans, at depths of 5000 m and more. There is,

■ THE SUN IS ABOUT TO DISAPPEAR BELOW THE HORIZON. A LUMINOUS HALO, CAUSED BY ATMOSPHERIC SCATTERING, SURROUNDS IT. THE DESERT TWILIGHT SEEMS STRANGELY FAMILIAR, BUT WE ARE WELL AND TRULY ON ANOTHER WORLD.

in the absolute darkness in the Movile Cave in Romania, a whole fauna that for thousands or even millions of years has been growing and becoming more complex without a single molecule of oxygen to breathe. So could there then be martian organisms dreaming away the time in hibernation in the freezing waters of some subterranean lake, far beneath the desert wind-swept plains? For 10, 100, or 1000 million years they may have been patiently waiting for better times, hidden in the subterranean ice of Sinus Meridiani, while the humans' robots search in vain for traces of life on the plains of Utopia and Chryse.

Phobos and Deimos:

pebbles in the sky

■ MARS IS ACCOMPANIED BY TWO SATELLITES: PHOBOS AND DEIMOS. THESE ARE TINY BODIES, COMPARABLE IN SIZE AND MASS TO SMALL ASTEROIDS, AND ORBIT JUST A FEW THOUSAND KILOMETRES FROM THE RED PLANET. HERE, PHOBOS APPEARS COVERED IN IMPACT CRATERS, LIKE THE MOON OR MERCURY. THE ORIGIN OF THESE SMALL CELESTIAL BODIES REMAINS UNKNOWN. ARE THEY ASTEROIDS CAPTURED BY THE PLANET MARS?

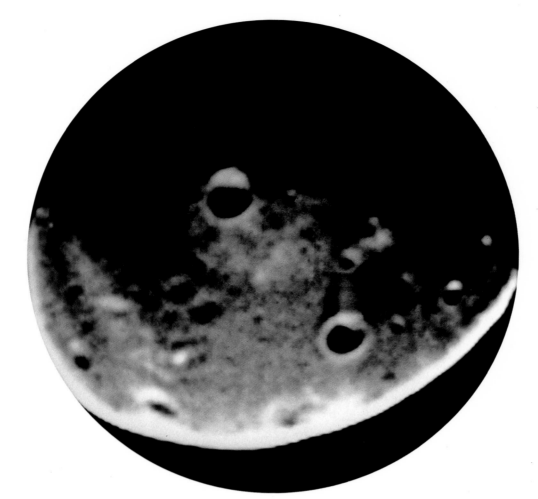

■ PHOBOS AND DEIMOS ARE EXTREMELY DARK BODIES. LIKE HUGE LUMPS OF COAL, THEY REFLECT ONLY ABOUT 5 PER CENT OF SUNLIGHT. DEIMOS, SHOWN HERE, HAS A SMOOTHER, MORE EVEN SURFACE THAN PHOBOS.

This plain, covered in very dark ashes and impact craters, resembles certain lunar landscapes. The black sky, dotted with stars, and dominated by a Sun of unchanging brilliance, is that of a body that does not have the slightest trace of atmosphere. We are perched on a hillock some 10 m high, which is a gigantic mountain on the scale that applies on this tiny world. From this height, the tortured landscape of Phobos may be seen as far as the horizon, which is very close. Here, our view is limited to just a few hundred metres. Beyond that, the curvature of the horizon is such that our line of sight stretches straight to the stars. After walking for just a few minutes on this tiny body, which brings the Little Prince's planet to mind, we discover a completely new landscape, and, after walking for two or three hours we find ourselves back at our starting point, having walked right round Phobos!

Yet Phobos offers one of the most beautiful vistas in the whole of the Solar System. Mars is right there, just above our heads, less than 6000 km away. Lifting our eyes towards the sky, we can see the red planet just as if we were flying above it in a spacecraft. From here, the gigantic network of canyons of Valles Marineris, wreathed in layers of cloud, is perfectly visible. Mars appears as a globe with an apparent diameter of 50°, one hundred times as large and as bright as the Full Moon seen from the surface of the Earth! Mars is so close that, even with the naked eye, tiny details are visible on the surface – craters less than 3 km in diameter, mountains swathed in high-altitude cloud, and even the water-cut valleys winding across the Utopia and Chryse plains.

Every 7^{h}30^m, Phobos completes one orbit of the planet, flying over the four great volcanoes of the Tharsis Ridge: Olympus, Arsia, Ascraeus, and Pavonis Mons. Seen from this distance, their immense size make one giddy. Might these sleeping, benign giants awaken one day?

Mars has two satellites: Phobos and Deimos. On the scale of objects in the Solar System, they are mere pebbles, which orbit at 9400 km and 23 500 km, respectively, from the centre of the planet. No other celestial body in the Solar System orbits so close to its parent planet as Phobos. Deimos takes slightly more than 30 hours to complete its orbit around Mars. Phobos and Deimos, unlike Mars, Earth, or the Moon, are not spherical in shape, but are instead extremely irregular triaxial ellipsoids. Their dimensions are 27×21×19 km, and 15×12×11 km, respectively. Their densities are 2.0 and 1.9. The mass of Phobos just exceeds 10 000 billion tonnes, while that of Deimos is less than 2000 billion. Such masses are, again on the scale of most objects in the Solar System, tiny. On Earth, it is very roughly the mass of Mount Everest. In fact, their gravitational attraction has never been high enough to cause them to collapse, be moulded and finally adopt the quasi-spherical shape of the major planets.

During the 21st century, astronauts will probably visit Phobos, which is an ideal viewpoint and outpost for observing the red planet. Even encased in spacesuits that give them a mass of more than 200 kg, those astronauts will weigh no more than about 200 grammes on Phobos! Exploration of the planetoid under those conditions, depending on the visitor's skill and training, will resemble a dream, or a nightmare. A slight push with the legs, and the explorers will – just as in a dream – be able to float up into Phobos' sky and, in a long, parabolic leap, cover a distance of as much as a hundred metres. But if their jump is too powerful, they run the risk of finding themselves hanging above the surface – even going into orbit! – for tens of minutes on end. Plenty of time for the astronauts, paradoxically trapped by the planetoid's feeble gravity, to admire the enormous red planet. Mars, seen from Phobos, remains absolutely fixed in the sky. The latter's major axis constantly points towards the planet. Throughout a day on Phobos, Mars passes through all its phases and, when it is

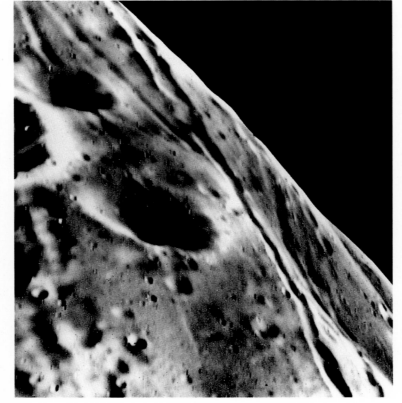

■ PHOBOS AND DEIMOS WERE SEEN BY THE VIKING PROBES IN MORE DETAIL THAN HAD EVER BEEN ACHIEVED UP TO THAT TIME BY ANY DISTANT SPACE MISSION. HERE, AN AREA MEASURING ABOUT 3 KM SQUARE SHOWS FURROWS, WHOSE ORIGIN IS STILL UNKNOWN, ON THE SURFACE OF PHOBOS. THEIR DEPTH IS SOME TWENTY METRES. THE SMALLEST CRATERS VISIBLE ON THIS IMAGE MEASURE NO MORE THAN 10 M ACROSS.

fully illuminated by the Sun, the strangest sight that it offers is perhaps that of the small, black, elongated patch, the shadow of Phobos, that appears to sweep across it from west to east.

Phobos, like the Moon or Mercury, is entirely covered with impact craters. The smallest, which measure just a few metres in diameter, are almost buried beneath the regolith, a thick layer of dust. The three largest craters, discovered at the time of the two Viking missions in 1976, have been named Roche, Hall, and Stickney. They measure 5, 6, and 9 km in diameter, respectively. Covering practically 10 per cent of the surface, the impact that created Stickney nearly disrupted Phobos completely. What is more, the larger of Mars' satellites also appears as if it had been worked by a gigantic plough, showing deep, long furrows. Perfectly aligned, 100 to 200 metres wide and some twenty-odd metres deep, they extend for several kilometres. Where do these bizarre grooves come from? Did the violence of the Stickney impact fracture Phobos in this fashion?

In its appearance, the surface of Deimos resembles that of Phobos. The same craters, the same hillocks, the same, almost black, dust that covers everything. Like the albedo of Phobos, that of Deimos is about 5 per cent. The surface of Deimos is, however, smoother, more regular. Here, there are no giant impact craters nor grooves. Deimos is probably covered with a thick layer of dust, which has softened the surface, disguising the shallowest craters, and masking minor relief. Here and there, however, rocks the size of houses emerge from the ash-grey plains.

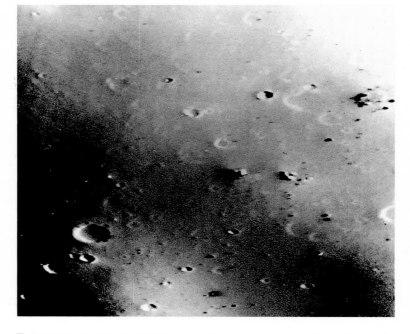

■ A VERY SMALL AREA OF DEIMOS, IN AN EXTREME CLOSE-UP FROM VIKING ORBITER 2. THE AREA IS NO MORE THAN 100 HECTARES! THE SMALLEST ROCKS VISIBLE ON THE SURFACE ARE JUST A FEW METRES IN SIZE. NUMEROUS IMPACT CRATERS APPEAR TO BE BURIED BENEATH A THICK LAYER OF METEORITIC DUST, SIMILAR TO WHAT ONE FINDS ON THE MOON.

Where do Phobos and Deimos originate? Mars is the only terrestrial planet to have two such mini-satellites. Neither Mercury nor Venus, nor even the Earth possesses anything similar. Phobos is undoubtedly too close to the planet to have been formed in situ by accretion. In fact, the satellite orbits inside the Roche Limit, which is a forbidden zone around a planet, where a body undergoes enormous gravitational tides. Inside the Roche Limit, interplanetary material cannot aggregate together, and has a tendency to fall towards the planet, or to become organised into rings, as has happened around the four giant planets. Almost all the experts nowadays agree that Phobos and Deimos are asteroids, which were captured by the red planet during the course of their travels through the Solar System. There are, however, serious objections to this apparently perfectly natural and reasonable theory. For example, the orbits of Phobos and Deimos are almost perfectly circular and lie almost exactly in the planet's equatorial plane. How can we explain such precise 'adjustments' to the orbits? No one is able to do so. One truly fantastic theory has been proposed to explain the existence of Phobos and Deimos: these two bodies are not asteroids, but are rocky bodies ejected by an immense eruption of one of the giant martian volcanoes! With heights up to 27 km and calderas 100 km in diameter, the volcanoes Arsia Mons, Ascraeus Mons, Olympus Mons, and

Pavonis Mons are all close to the martian equator, i.e., directly beneath Phobos and Deimos, and are thus ideal suspects. But the energy required to insert bodies as large as Phobos and Deimos into orbit is colossal, one hundred and fifty times as great as the most violent eruption ever recorded on Earth. So is it science fiction?

Whatever its mode of formation, Phobos is too close to the martian surface for it to remain there for ever and a day. The immense mass of Mars is inexorably drawing Phobos towards it. Braked in its orbit by gravitational tides, Phobos is inexorably being slowly drawn down towards Mars. In about 30 million years, a mountain will fall out of the sky and crash onto the red planet, excavating a new crater, raising a gigantic storm of dust in the martian atmosphere that will last for centuries, plunging the planet into darkness and a terrible glaciation. Whilst waiting for this distant planetary catastrophe, researchers have time to investigate the secrets of Phobos and Deimos: the origin of these two chunks of rock that hurtle across the martian sky is still unknown.

Gaspra:

our first asteroid

■ THE GREAT FAMILY OF ASTEROIDS CURRENTLY INCLUDES
MORE THAN FIVE THOUSAND MEMBERS, BUT
ASTRONOMERS SUSPECT THAT THERE ARE MILLIONS OF
THESE SMALL BODIES IN THE SOLAR SYSTEM. AMONG ALL
THESE BODIES, GASPRA WAS THE FIRST TO BE
APPROACHED AND STUDIED FROM ITS IMMEDIATE
VICINITY BY A SPACEPROBE. THE SMALLEST
ASTEROIDS MEASURE JUST A FEW METRES IN
DIAMETER. THE LARGEST, CERES, IS ALMOST
ONE THOUSAND KILOMETRES ACROSS.

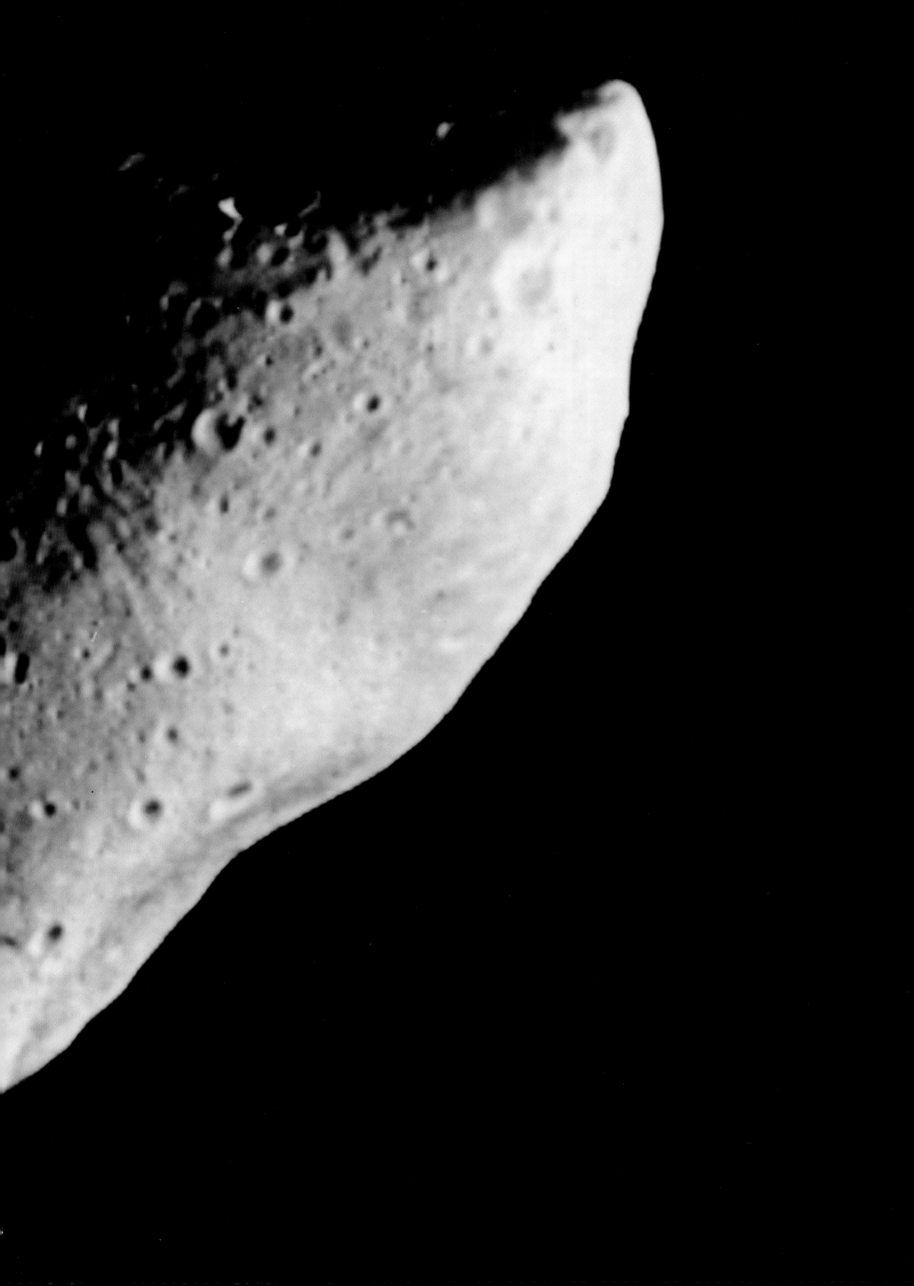

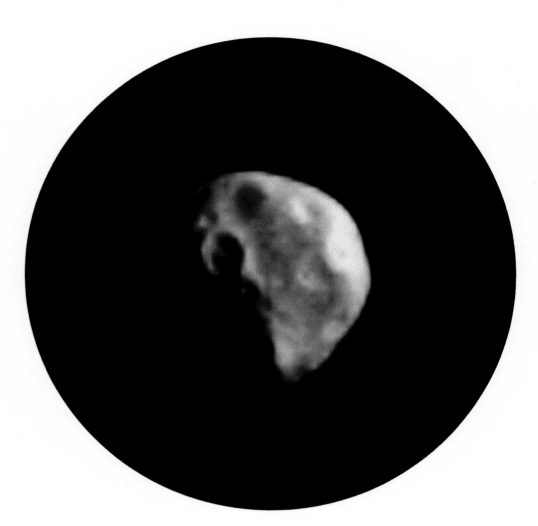

■ DACTYL IS ONE OF THE TINIEST OF KNOWN OBJECTS IN THE SOLAR SYSTEM. THIS SATELLITE OF THE ASTEROID IDA MEASURES JUST 1500 M IN DIAMETER! THE LARGEST CRATER VISIBLE ON THE TERMINATOR IS 300 M ACROSS, AND THE PHOTOGRAPHIC RESOLUTION OF THIS IMAGE IS ABOUT 40 M.

Silence and darkness. In the sky, there is no moon, no giant planet displaying its glowing polar aurorae or its swirls of gas driven by storm-force winds. An empty space, with no other points of reference than the cold, distant stars of the Milky Way. A frightening, austere place. The landscape that surrounds us is utterly invisible, and it is only the absence of stars below the horizon that shows that we are not floating in interstellar space. The force of gravity of the world on which we stand is ridiculous. An astronaut here would weigh less than 100 grammes, and could escape from the feeble gravitational field with an energetic jump. After three-and-a-half hours of total darkness, the Sun finally rises above the horizon of this airless world. Seen from here, the Sun's disk is minute, one tenth of the brightness of the warm Sun on Earth. The first rays of this extraterrestrial dawn illuminate a chaotic field of low hills and impact craters that here and there interrupt a dismal plain of brownish ash. The horizon is just a few tens of metres away.

This rock seems to be lost in utterly empty space, and yet we are on Gaspra, in the most densely populated region of the Solar System! Between Mars and Jupiter lies the asteroid belt,

a vast torus which extends, in rough figures, between 300 and 600 million kilometres from the Sun, and where millions, or even billions of rocks of all sizes have their orbits.

Gaspra would have remained an obscure, anonymous chunk of rock if it had not been the first asteroid to be visited by a spaceprobe from Earth. On 29 October 1991, the American Galileo probe, which was *en route* to Jupiter, passed some 1600 km from this small body, orbiting 330 million kilometres from the Sun: a really close fly-by. In the few hours of the fly-past, Galileo's powerful telescope showed us a new world. Until then, it had never actually been possible to observe asteroids, too small and too far from Earth, in any detail. To astronomers they had never been more than tiny points of light in the sky.

Gaspra is a large, irregular, rocky body, measuring $19 \times 12 \times 11$ km, and with a mass of some 5000 billion tonnes. Its size lies between those of Phobos and Deimos and, like them, it is covered in a thick layer of regolith, which is itself peppered with impact craters. But, unlike Mercury, the Moon, or Phobos, Gaspra's dusty plains are not saturated with impact craters. The Galileo images reveal less than a hundred, the

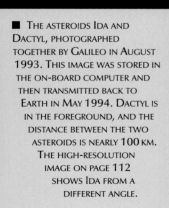

■ The asteroids Ida and Dactyl, photographed together by Galileo in August 1993. This image was stored in the on-board computer and then transmitted back to Earth in May 1994. Dactyl is in the foreground, and the distance between the two asteroids is nearly 100 km. The high-resolution image on page 112 shows Ida from a different angle.

largest of which do not exceed 1500 m in diameter. This discovery is surprising and very important. It means that, unlike the planets and most of their satellites, Gaspra must be a young object. The surface area occupied by its craters and their sizes suggest an age of just about 200 million years! Gaspra's silhouette, dubbed the 'shark's mouth', may perhaps offer a solution to this enigma. The small asteroid probably consists of two blocks – two ancient asteroids – that collided during what was the Mesozoic era on Earth. Mutual collisions seem to be the fate of many asteroids. Castalia, a mini-asteroid 2 km in size, discovered in 1989 not far from Earth, also consists of two large pieces in contact, like Toutatis (also discovered in 1989) which made a close approach to Earth late in 1992, when telescopes revealed that it was double. Finally, the Galileo probe encountered another asteroid, Ida, in August 1993. Ida measures 52 × 30 × 24 km and has a satellite, Dactyl, 100 km distant and measuring just 1.5 km in diameter.

Early in 2002, twenty-odd asteroids have known satellites, including 22 Kaliope, 45 Eugenia, 617 Patroclus, 762 Pulcova, and 1998 WW$_{31}$. Pulcova, which measures 150 km in diameter, has a small 15-km satellite, just 800 km distant. Eugenia is a large asteroid known since 1857, measuring 215 km across, and orbiting between Mars and Jupiter at a distance of 400 million kilometres from the Sun. It is accompanied by a mini-asteroid, 13 km across, which orbits at 1200 km in slightly less than 5 days.

Although Gaspra and Ida are now the most famous and best known of all the asteroids, they actually occupy a fairly modest place among the minor planets. Hundreds of their fellow bodies are larger and more massive. Researchers estimate that several billions of tiny asteroids, just a few metres in diameter, exist in the Solar System. Every year, hundreds of these rocks fall on the Earth after having partially broken up in penetrating the atmosphere at several thousands of kilometres per hour. These meteorites are found more or less all over the globe, and have wandered between the planets for billions of years before encountering the Earth. The most important asteroids,

■ ON 28 AUGUST 1993, THE GALILEO SPACEPROBE PASSED LESS THAN 2500 KM FROM THE ASTEROID IDA, AND SENT THIS ELECTRONIC IMAGE BACK TO EARTH, SHOWING A VERY ELONGATED ROCKY BODY, MEASURING 52 × 30 × 24 KM. IDA, TWICE THE SIZE OF GASPRA, ALSO SEEMS TO BE OLDER, AS SHOWN BY THE NUMEROUS CRATERS. SOME OF THE IMPACTS APPEAR TO BE GREATLY ERODED.

measuring at least 1 km in diameter, probably amount themselves to about one million – the rest remain to be discovered. At present, astronomers possess full information about more than 5000 asteroids, which, like Castalia, Toutatis, or Gaspra, measure between 1 and 20 km in diameter. But the largest of them far exceed Gaspra's size. About one hundred asteroids measure more than 100 km in diameter. Like Gaspra or Phobos, these objects are not massive enough to have the spherical shape that gravity creates in the planets and the larger satellites. Probably ellipsoidal in shape, Vesta and Pallas are more than 500 km in diameter. Finally, the largest of the Solar-System asteroids, Ceres, is a remarkable exception. It is a proper small planet, spherical in shape, that is almost 1000 km in diameter. Ceres must resemble the satellites Tethys, Dione, Ariel, Umbriel, and Charon, which are all almost exactly the same size. But these other objects primarily consist of ice, whereas Ceres is a denser body that is mainly rock. With its mass that exceeds one billion billion tonnes, Ceres represents perhaps one quarter of the total mass of all the billions of asteroids that orbit the Sun! It takes four-and-a-half years to complete its orbit, which is slightly elliptical, at an average distance from the Sun of 414 million kilometres.

Researchers believe that when the Solar System originated, Ceres, like Venus, the Earth, and Mars, started to become a larger planet by sweeping up material from the disk that surrounded the young Sun. It was the proximity of the largest planet in the Solar System, Jupiter, that prevented this process from continuing. Jupiter's enormous mass would have perturbed the orbits of the asteroids, preventing them from accreting into a single object. The objects in what came to be known as the asteroid belt are, in some respects, the Solar System's fossils, showing us an almost intact remnant of the original protoplanetary disk. Some bodies, such as Gaspra, have been born recently, from the collision of two more ancient planetoids, and others are probably intact, having orbited the Sun, unchanged, for 4.5 billion years. Over the course of time, the asteroids have been forced to alter their

■ THE ASTEROID MATHILDE, OBSERVED IN DETAIL FOR THE FIRST TIME BY THE NEAR SPACEPROBE, WHEN *EN ROUTE* FOR ITS RENDEZVOUS WITH EROS. THIS ASTEROID, MEASURING 66 × 48 × 46 KM, AND ORBITING AT A DISTANCE OF 300 MILLION KILOMETRES FROM THE SUN, HAS BEEN EXCAVATED BY A DEEP CRATER. NEAR'S FLY-PAST OF MATHILDE TOOK PLACE ON 27 JUNE 1997, AT A DISTANCE OF LESS THAN 1800 KM.

orbits by the gravitational forces exerted by the major planets, and have become clustered in the large families consisting of several thousand objects, which orbit in denser zones. These regions are separated by empty spaces, known as 'gaps', in which gravitational resonances cause orbits to be unstable, and from which they have expelled the asteroids.

After the success of the Galileo mission, which revealed Gaspra, Ida, and Dactyl, the closing years of the 20th century saw planetologists become more and more enthusiastic about these fragmentary planets. Several space missions attempted, with greater or lesser success, to fly past an asteroid. For the Near (Near-Earth Asteroid Rendezvous) mission, the trip was fraught with pitfalls. The small American probe left Earth on 17 February 1996 for Mathilde, a wonderful asteroid measuring 66 × 48 × 46 km. The encounter occurred without a hitch on 27 June 1997 at a distance of

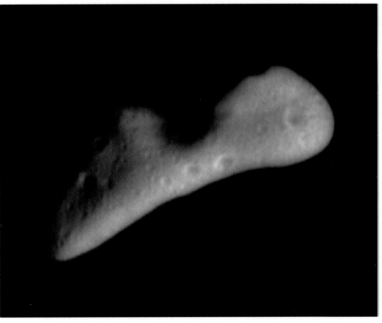

■ THE SMALL AMERICAN NEAR SPACEPROBE TOOK THIS FINE COLOUR IMAGE AS IT APPROACHED EROS IN FEBRUARY 2000. THE ASTEROID MEASURES 40 × 14 × 14 KM. THE MASS OF EROS, MEASURED PRECISELY THANKS TO VARIATIONS IN THE PROBE'S PATH, WHICH WAS GREATLY PERTURBED BY ITS IRREGULAR GRAVITATIONAL FIELD, IS ABOUT 6700 BILLION TONNES.

less than 1800 km. During its fly-by, Near photographed Mathilde from every angle, revealing impressive impact craters, some of which measured 20 km in diameter. Then Near set course for another asteroid, Eros, but unfortunately the probe missed the rendezvous. On 10 January 1999, a fault in the probe's rocket prevented it from being put into orbit around Eros, as intended. However, the American engineers took brilliant control of Near and, for more than a year, kept their spacecraft on an orbit close to that of Eros, some 320 million kilometres away from Earth, so that they could attempt a second approach in February 2000. This succeeded perfectly and allowed planetologists to obtain spectacular images of the surface of the asteroid, which measures 40 × 14 × 14 km.

Encouraged by the success of their mission, the American astronomers then decided to end it with a flourish, by landing Near on Eros! Gradually, starting in January 2001, the probe

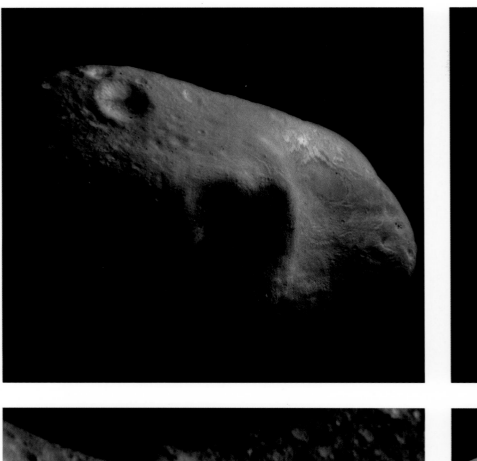

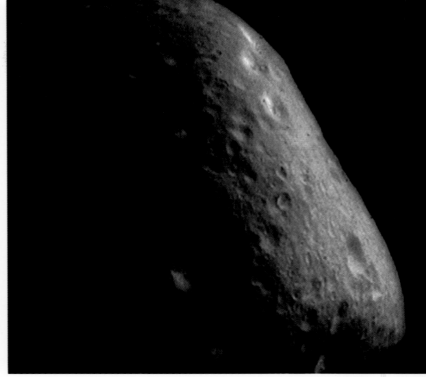

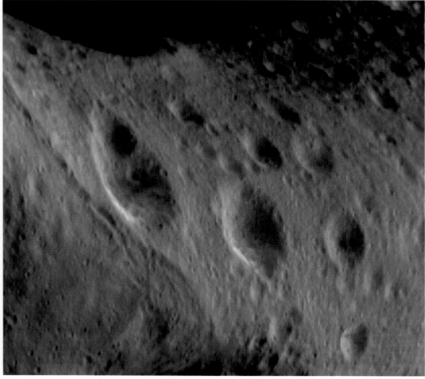

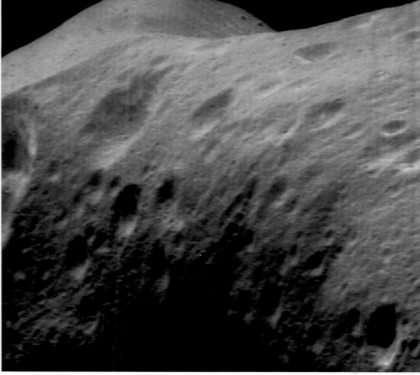

approached the asteroid, finally falling onto it on 12 February. During its final

on either side of the giant planet. Chiron, discovered in 1977, has a rather unusual

descent, Near continued to transmit images of the surface with increasing resolution. On the very last images, taken at a height of about one hundred metres, the ground of Eros appears to be strewn with stones, a few tens of centimetres across. Apart from the notable exceptions of the Moon, Venus, and Mars, no other celestial body has been seen from so close. Once it arrived at the surface, where it must have bounced several times before coming to rest, the probe fell silent. The performance that the NASA engineers achieved in remote-control piloting the probe is remarkable, because it was never originally expected to end its career on the surface of Eros.

Although 95 per cent of the asteroids orbit, like Gaspra or Ceres, between Mars and Jupiter, some have very different orbits. For example, the asteroids in the Trojan group, trapped by Jupiter's gravity, precede or follow the latter in its orbit, 60°

orbit, which takes it from the vicinity of Saturn to beyond Uranus. The object 1992 QB$_1$, discovered at the end of 1992, and which has yet to be named, currently holds the distance record, because it lies beyond Pluto, at about 9 billion kilometres from the Sun! By contrast, some small asteroids come very close to the Sun, like Icarus, which ventures inside the orbit of Mercury. The Amor group of asteroids crosses the orbit of Mars. Phobos and Deimos may perhaps have been captured by the red planet during a close approach. The asteroids in the Apollo group are objects 1 to 5 km in size that venture across the Earth's orbit. Apollo, Adonis, Hermes, Geographos, Icarus and, more recently, Toutatis, have all flown very close to our planet in the past. Over a period of millions of years one of these mountainous lumps of rock will undoubtedly crash into the Earth, as asteroids have done

regularly since the formation of the Solar System. Currently, astronomers are trying to

take place. In the Arizona desert, for example, the famous Meteor Crater, a

establish a network for monitoring asteroids, to determine those that might be dangerous for our planet. These disturbing objects, known as 'Earth-crossers' to the specialists, may attain quite respectable dimensions. Eros measures 40 km in diameter; Sisyphus, 8 km; Ivar, 8 km; Tezcatlipoca, 6 km; and Hephaistos and Toutatis about 5 km. What all have in common is a perihelion distance – closest approach to the Sun – similar to that of the Earth: 148 million kilometres for Ivar, for example. This systematic surveillance is beginning to bear fruit. On 9 December 1994, for example, just fourteen hours after its discovery, the asteroid 1994 XM_1, a rock some fifteen metres in diameter, passed just 104 700 km from the Earth, which, to date, remains a record. Or rather, it is not. The dozens of craters discovered by geologists on our planet bear witness that, from time to time, collisions do well and truly

bowl some 1200 m across and 200 m deep, was excavated by a rock of just 20 to 30 m in diameter some thirty thousand years ago.

At the Cretaceous–Tertiary boundary, 65 million years ago, the fall of an asteroid some 10 km across probably caused the disappearance of 75 per cent of all animal and plant species. Today, the impact of Icarus, Toutatis, or a similar body would utterly disrupt the biosphere. Our fragile species might not survive the havoc wreaked by the impact and the long, terrible period of glaciation that would follow. While we wait for Armageddon, astronomers, armed with ever more powerful telescopes, currently detect several tens of asteroids every year. But the vast majority of asteroids remain to be discovered, and some undoubtedly have orbits that intersect that of the Earth.

Jupiter:

the planet of storms

■ JUPITER, THE LARGEST PLANET IN THE SOLAR SYSTEM, MORE THAN THREE HUNDRED TIMES THE MASS OF THE EARTH, IS BASICALLY A GASEOUS WORLD, WHERE STORMS RAGE THAT ARE CAPABLE OF ENGULFING ENTIRE PLANETS. THIS PHOTOGRAPH SHOWS AN AMAZING VIEW OF THE GIANT PLANET. IN THE FOREGROUND, AGAINST A BACKDROP OF THE PLANET'S BRILLIANTLY COLOURED CLOUDS, ARE TWO OF THE LARGEST SATELLITES, IO (LEFT) AND EUROPA. EACH OF THESE BODIES IS LARGER THAN THE MOON.

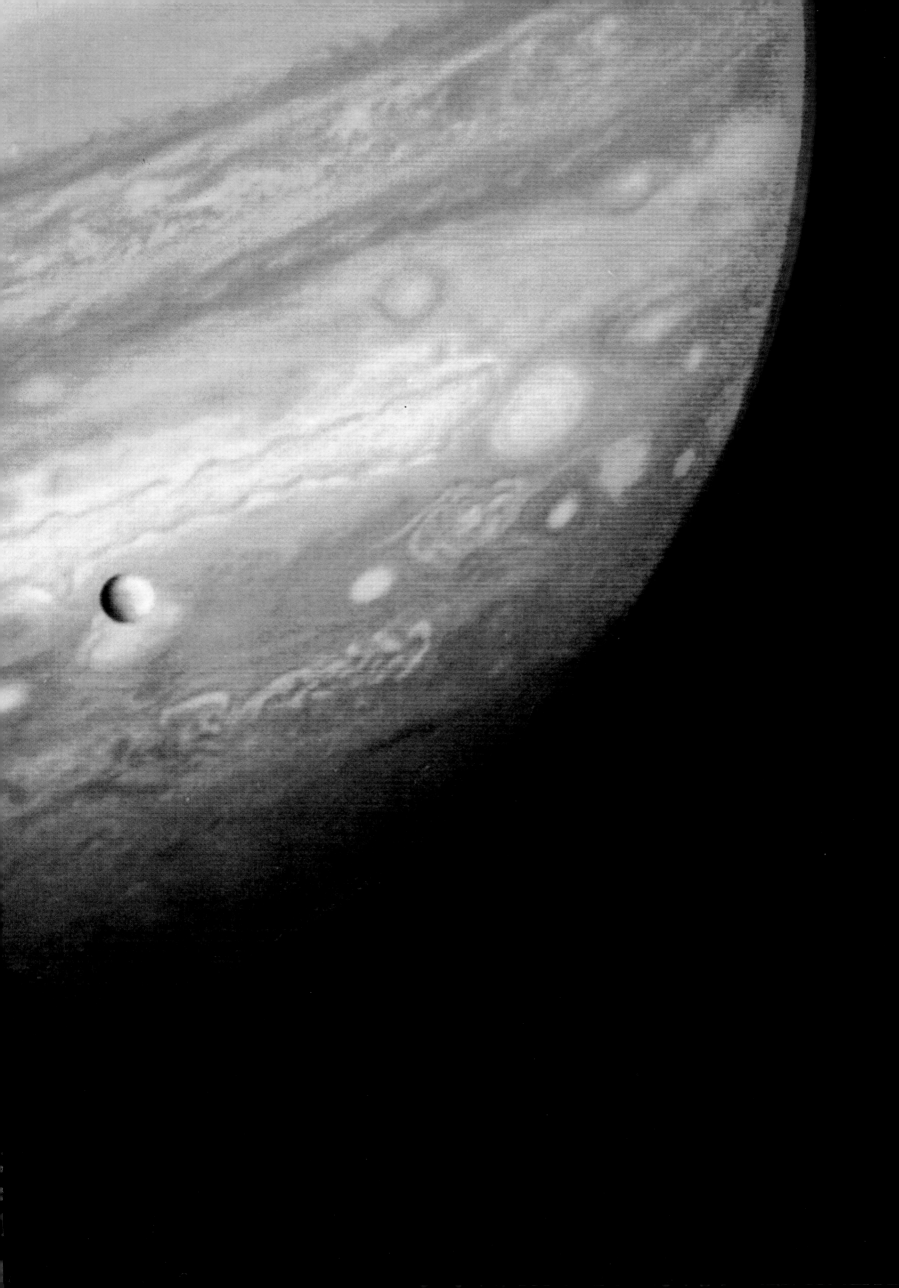

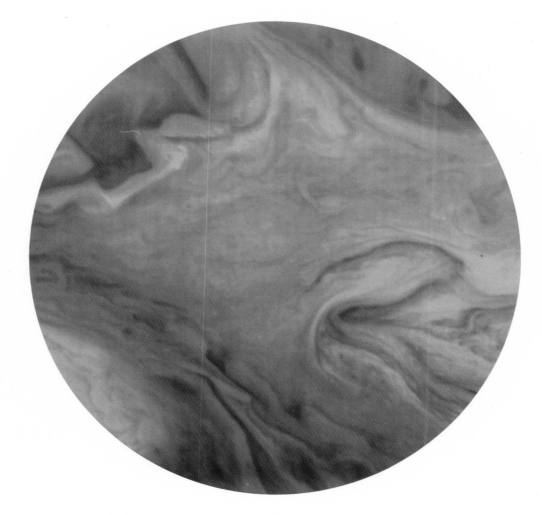

■ THE UPPER ATMOSPHERE OF JUPITER, OBSERVED IN CLOSE-UP BY THE VOYAGER 1 SPACEPROBE. THE ORIGIN OF THE COLOURS IN THESE CLOUDS IS STILL NOT FULLY ESTABLISHED. THE FINEST DETAILS VISIBLE HERE ACTUALLY MEASURE ABOUT 30 KM ACROSS.

Curls of ochre and yellow clouds flying silently past occasionally hide the blue sky. The wind is blowing a gale, and the Sun, visible between two clouds, appears minute and far away. It is cold, −120°C, but the atmospheric pressure is the same as on Earth. The air, bitterly cold and almost completely devoid of oxygen, is unbreatheable. Any astronaut trapped here on this storm-wracked planet, would be hurled around by the gusts of wind like a wisp of straw in a cyclone. Impossible to grab hold of a rock, because here there is no ground and we are up in the air, among the clouds of Jupiter, the largest of all the planets. Wind and clouds are the rulers of this featureless world. Above us, there are bright ammonia-ice cirrus clouds, racing past at a breakneck speed, and the ochrous mountains that tower over us continue to grow but are torn apart by the force of the wind, which is blowing at more than 500 km/h.

Beyond the five terrestrial planets – whose surfaces may sometimes be wild and hostile, sometimes pleasant and reassuring, but with are always of human dimensions – the Solar System exhibits a series of other worlds that are radically different. Often strange, frightening, or even utterly inhuman, these bodies are dominated by the four giant planets, spheres of gas and the site of unbridled atmospheric disturbances. Here, supersonic winds raise hurricanes in which the Earth itself would be lost. Worlds of gases and clouds, Jupiter, Saturn, Uranus, and Neptune contain within the four of them, more than 99 per cent of the mass of all the planets in the Solar System. Around them orbit several dozen small worlds, ranging from tiny lumps of rock to planet-sized bodies of rock and ice.

Jupiter is the largest of the giant planets. It has an equatorial diameter of some 142 000 km, and completes its orbit, at 778 million kilometres from the Sun, in slightly more than eleven Earth years. The mass of Jupiter is about 2 million billion billion tonnes, which is 318 times that of our small Earth. Although gigantic in terms of its small planetary siblings, Jupiter is, nevertheless, tiny when compared with the Sun, which is more than one thousand times as massive as Jupiter!

Despite its enormous mass, Jupiter is basically a gaseous world. Its density of 1.34 is practically identical to that of the Sun and, like our star, the giant planet consists of 98 per cent hydrogen and helium, and about 2 per cent of more complex

■ THE UPPER ATMOSPHERE OF
JUPITER CONTAINS CYCLONIC
SYSTEMS THAT MAY PERSIST FOR
YEARS, OR EVEN CENTURIES. THIS
IMAGE SHOWS THE SOUTH
TROPICAL ZONE AT THE TOP,
DOMINATED BY THE FAMOUS
GREAT RED SPOT. IN THE
CENTRE OF THE IMAGE, THE
ANTICYCLONE FORMING THE
BRIGHT WHITE OVAL IS AS
LARGE AS THE EARTH.

■ LATE AFTERNOON ON JUPITER. THE GREAT RED SPOT IS NEARING THE GIANT PLANET'S LIMB AND WILL SOON BE PLUNGED INTO DARKNESS. ON THIS IMAGE, THE ATMOSPHERIC DISTURBANCES CREATED BY THE SOLAR SYSTEM'S LARGEST STORM ARE VISIBLE. TO THE SOUTH, A STREAM OF GAS SNAKES FROM THE GREAT RED SPOT TO A WHITE OVAL. TO THE NORTH, THE WHOLE OF THE CLOUD BAND IN WHICH THE RED SPOT LIES APPEARS TO BE DISTURBED FOR TENS OF THOUSANDS OF KILOMETRES.

trace elements such as methane, ethane, ammonia and acetylene. One thousand three hundred times the volume of the Earth, Jupiter rotates on its axis in less than ten hours. Such a rate of rotation, linked with the planet's enormous mass and its low density produces a significant centrifugal effect, which expands its equator and flattens it at the poles. As a result, Jupiter is not spherical but is markedly ellipsoidal, with the polar diameter, 133 000 km, being significantly less (by almost the diameter of the Earth!) than the equatorial diameter, 142 000 km. The shape of the giant planet and its high rate of rotation are the origin of its very notable atmospheric conditions, where bands of cloud, parallel with the equator, slide past one another. As in the terrestrial atmosphere, cyclonic and anticyclonic features develop along the boundaries between the various cloud bands. The largest of these features reach such sizes that they are easily visible from Earth. Moving in Jupiter's upper atmosphere, varying in size, definition, and colour, these phenomena of extraordinary size are continually followed from Earth not only by professional observers, but also (and primarily) by amateur astronomers, who are able to study them with very small telescopes.

But back to Jupiter's upper atmosphere, where four crescent moons are visible in the deep blue sky. The closest of these moons, Io, a vivid yellow colour, seems larger and brighter than the Moon when seen from Earth. In the far distance, towards the horizon, which is hundreds of kilometres away, we can make out an enormous, deep red wave in the clouds. Seen from such a great distance, this gigantic, disturbing wave seems stationary. It is in fact the outer edge of the most extraordinary cyclone in the whole Solar System. Swirling for hundreds of years in Jupiter's atmosphere, the Great Red Spot is a monstrous maelstrom, 30 000 km in diameter, rotating around its centre every

■ THIS FAMILIAR IMAGE OF JUPITER WAS OBTAINED FROM EARTH ORBIT BY THE HUBBLE SPACE TELESCOPE ON 28 MAY 1991. NOTE THAT THE PHOTOGRAPH OPPOSITE WAS TAKEN BY VOYAGER 2 TWELVE YEARS EARLIER, ON 29 JUNE 1979. THE HUBBLE IMAGE CLEARLY SHOWS THE PARALLEL ATMOSPHERIC BELTS THAT ENCIRCLE THE PLANET, AS WELL AS THE GREAT RED SPOT.

■ FAR LESS SPECTACULAR THAN THE RINGS OF SATURN, THOSE OF JUPITER ARE DIFFICULT TO OBSERVE. THEY ARE BACK-LIT HERE, AT THE RIGHT, WHILE JUPITER'S UPPER ATMOSPHERE APPEARS AS A NARROW COLOURED CRESCENT.

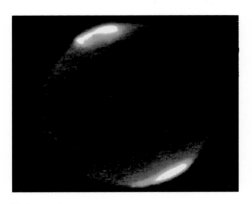

■ JUPITER'S MAGNETIC FIELD IS TEN TIMES AS STRONG AS OUR OWN PLANET'S. THE FIELD IONIZES ATOMS IN JUPITER'S UPPER ATMOSPHERE, SURROUNDING BOTH POLES WITH GLOWING AURORAE.

six days and capable of swallowing the Earth several times over. The Great Red Spot rears its enormous phosphorus cumulus clouds 20 km above the general level of the jovian atmosphere and leaves behind it a turbulent wake several thousand kilometres long. Nothing here is on a terrestrial scale. Yet Jupiter's upper atmosphere, with its blue sky, its pressure and temperature almost comparable with those on Earth, might seem almost familiar, if the clouds that surround us were not as large as planets, and did not hide an ocean, so dark and so deep as to give us vertigo. A dive deeper into the layers of clouds would reveal a completely different world, totally hostile to the materials that we find on our tiny blue planet. By the end of the 20th century, humans had sent five ambassadors to Jupiter. Five probes, all American, which, in less than twenty-five years completely overturned our knowledge of the largest of the planets. These were, first, the Pioneer 11 and 12 probes, which left the Earth on 2 March 1972 and 5 April 1973, respectively, and which reached the orbit of Jupiter on 4 December 1973 and 2 December 1974. The instruments on board the small Pioneer probes, which weighed no more than 250 kg each, revealed the strength and complexity of Jupiter's magnetic field and provided the first high-resolution images of the planet. The results obtained by the next pair of probes, the famous Voyager 1 and 2 spacecraft, however, were on a completely different scale. Much larger – the Voyagers each weighed 815 kg – much more sophisticated, equipped with an onboard computer and a high data-rate transmission system – one hundred times as powerful as the Pioneer equipment – and two high-resolution cameras, the two American spaceprobes returned, in just a few months, ten times as much data as had been patiently accumulated ever since the first telescopic observation of Jupiter by Galileo in 1609.

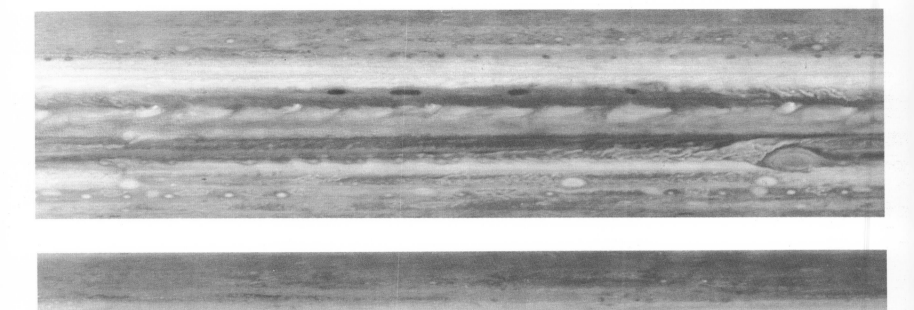

■ THESE TWO STRIP CHARTS OF JUPITER WERE PREPARED FROM DATA RECORDED BY THE VOYAGER PROBES. VOYAGER 1 *(TOP)* ENCOUNTERED THE PLANET AT THE BEGINNING OF FEBRUARY 1979. VOYAGER 2 *(BOTTOM)* FLEW PAST JUPITER AT THE END OF JUNE 1979. IN JUST OVER FIVE MONTHS MANY OF THE CLOUD SYSTEMS HAD CHANGED OR MOVED. THESE CHARTS MAY BE COMPARED WITH THE HUBBLE IMAGE SHOWN ON PAGE 121.

Voyager 1 was launched on 5 September 1977, on board a Titan-Centaur rocket, which lifted it out of the Earth's gravitational field at a velocity of 50 000 km/h. After less then ten hours, the probe crossed the orbit of the Moon, and it then plunged into the Solar System, heading for Jupiter, which it reached on 5 March 1979. Voyager 2 had left a few days earlier, on 20 August 1977, but crossed Jupiter's orbit four months later than Voyager 1, on 9 July 1979. It is to the two Voyagers that we owe the most extraordinary views of the Solar System ever taken in the history of spaceflight, not just dizzying zooms across the cyclones raging on the giant planet, but also the images of the four large Galilean satellites that are truly like something out of science fiction. Millions of numerical data points, and thousands of electronic images in colour were transmitted back to the Earth by the two probes over the course of a few months. Voyager 2 then successfully continued its mission to explore the Solar System, first visiting Saturn, then Uranus and finally Neptune. As far as Voyager 1 is concerned, it has long since passed the orbits of the two outer planets, Neptune and Pluto. At the beginning of the year 2000 it was more than 11.2 billion kilometres away, continuing with its endless odyssey, and still in contact with the Earth, via NASA's giant parabolic antennae.

■ EQUIPPED WITH A NEW PLANETARY CAMERA AT THE END OF 1993, THE HUBBLE SPACE TELESCOPE IS NOW ABLE TO PERMANENTLY FOLLOW THE EVOLUTION OF JUPITER'S ATMOSPHERE. THIS IMAGE, COMPARABLE WITH THOSE FROM THE VOYAGER SPACEPROBES, WAS TAKEN ON 18 MAY 1994. IO AND ITS SHADOW ARE SEEN CROSSING THE DISK OF THE PLANET.

The Galileo spaceprobe was given the perilous task of following the celebrated Voyagers. This impressive machine, weighing 2200 kg, carried into orbit by the Space Shuttle Atlantis, left Earth on 18 October 1989. Before reaching Jupiter's system, Galileo encountered the two asteroids Gaspra and Ida, which it photographed whilst passing. It was, however, a spaceprobe in a rather pitiable state that reached the giant planet on 7 December 1995. In fact, its large parabolic antenna, nearly 5 m in diameter, which should have enabled it to send back an unprecedented amount of data, primarily from its powerful telescope and its wide-field camera, never deployed properly. A small reserve antenna and the transmission of new programs to the on-board computer for compressing the data did, however, enable a small fraction of the mission to be rescued. The data recorded by the probe were transmitted back to Earth at about the same rate as that given by the old Pioneer probes. The Galileo mission was twofold. First, on its closest approach to Jupiter, it released a 340-kg module that descended into the atmosphere. Once that had been done, Galileo was put into orbit around Jupiter, where it successfully completed nearly thirty elegant elliptical orbits, like the petals of a flower, approaching the four large satellites of Jupiter one

■ On 7 December 1995, a probe released by the Galileo spacecraft plunged into the atmosphere of a giant planet for the first time. With breathtaking rapidity, the Galileo probe reached a depth of 160 km before suddenly ceasing its transmissions. Even lower, it was probably crushed by the atmospheric pressure.

after the other. After five years of continuous observations, Galileo's mission to explore Jupiter's system came to an end in 2000, after it had provided incredibly detailed images of the jovian satellites.

But we should not forget that on 7 December 1995, the small scientific module released by Galileo allowed us for the first time – albeit through the eyes of a robot – to dive deep into the clouds of the Solar System's largest planet. The probe entered the upper atmosphere at the highest velocity ever attained in spaceflight: more than 170 000 km/h!

During its dizzying descent into the yellow and white clouds, like every craft entering the atmosphere of any planet, the probe first had to survive. Protected by its thermal shield and heated to more than 15 000 °C, it was subjected to an incredibly violent deceleration, which no other craft had ever experienced: more than 200g! Only subsequently was its parachute deployed, enabling it to descend into the jovian clouds at a more reasonable speed. As it descended towards the centre of the planet, without hope of return, the Galileo descent module measured the atmosphere's increasing temperature, which reached 0 °C at a pressure of 5 bars. Lower still, at a depth of about 50 km, in a region inaccessible to terrestrial telescopes, the probe slowly passed through clouds where the temperature was about 30 °C at a pressure that was similar to that found at a depth of about 70 m in an ocean on Earth. So far, everything had gone well, because a probe is perfectly capable of surviving such an environment.

Perhaps future astronaut-divers could do the same?

Here, the sunlight, diffused by the dense clouds, resembles our own twilight. Is it possible that, among these dense clouds of hydrogen and helium, and also of water vapour, strange diaphanous wind-sailors glide silently and indifferently along, drifting at the whim of the winds? And did they see the Galileo probe drop through their aerial world, swinging gently beneath its parachute?

It was there, in the half-light of the turbulent clouds, that the Galileo probe made its most surprising discovery. The atmosphere of the giant planet contains abundant signs of gases that no one expected to find there: argon, krypton, and xenon. These three rare gases do not enter into combinations with other elements, and physicists know that they cannot have been captured by the planet except under extremely cold conditions, about −130 °C, which is a temperature lower than that found on the surface of Pluto. When, and how, did these gases arrive in the jovian clouds? According to planetologists, it could well be that Jupiter was formed, some 4.5 billion years ago, very far from the Sun, probably well beyond the orbit of Pluto, between 5 and 10 billion kilometres from the Sun. In the relatively dense and viscous protoplanetary disk, the nascent giant planet would have slowly spiralled in towards the Sun, before finally becoming stabilized. Once the gas and dust in the disk had condensed and accreted, the planet settled into an independent circular orbit. This scenario, which would have appeared utterly fantastic to the most serious experts on the

■ JUPITER, THE MOST MASSIVE AND LARGEST PLANET IN THE SOLAR SYSTEM PROBABLY HAS HUNDREDS OF MILLIONS OF COUNTERPARTS IN OUR GALAXY. SINCE 1995, ASTRONOMERS HAVE DISCOVERED DOZENS OF PLANETS IN ORBIT AROUND OTHER STARS, IDENTICAL TO THE SUN. MOST OF THE PLANETS SEEM TO HAVE CHARACTERISTICS SIMILAR TO THOSE OF JUPITER.

Solar System just a few years ago, has been supported by the recent discovery of giant planets around other stars. These objects, with masses comparable with that of Jupiter, often orbit at very small distances from their stars, sometime just a few tens of million kilometres. To explain the presence of massive bodies so close to stars, specialists have again had to postulate a slow process of migration.

The Galileo probe, continued its breakneck plunge towards the dark depths of the centre of Jupiter. Lower still, the temperature of the giant planet increases. Beneath the clouds, at a depth of about 100 km and a pressure of 10 terrestrial atmospheres, it exceeded 100 °C; here the Sun is invisible, and the heat is that produced by the planet itself. Jupiter, uniquely in the Solar System, actually emits nearly twice as much energy as it receives from the Sun. In the darkness of the jovian clouds, occasionally split by blindingly bright flashes of lightning, the tiny module from Earth sank into an atmosphere that became denser and denser. Soon it was lost in inky blackness and subjected to terrible pressure and temperature. After sinking for an hour, at a depth of about 160 km, in gas heated to 150 °C, and at a pressure of 24 bars, the Galileo probe finally fell silent.

Lower still (because we have still hardly scratched the cloud blanket that covers Jupiter's immense globe), the dreadful values that apply at the surface of Venus, 90 bars and 460 °C, are exceeded. Galileo hardly survived that far. At a depth of a few hundred kilometres, lost, crushed, and vaporized, it was not capable of measuring temperatures of thousands of degrees, nor pressure of thousands of bars. The end of the journey may only be imagined: who, or what, could survive such fantastic physical conditions? At depths of a few thousand kilometres, the numbers become mere abstract concepts, and the state of the matter itself becomes difficult to describe. What the coloured clouds of Jupiter hide from our view is a sort of planetary ocean, perhaps gently rocked by large-amplitude waves, and crushed beneath a pressure of millions of atmospheres. Even lower, all the gases become 'metallic' fluids, with strange properties, and where nothing moves. At a depth of 57 000 km we have reached the giant planet's rocky core.

This, still poorly known, is thought to be twice the size and about ten times as massive as the Earth. According to the planetologists, Jupiter must have originally formed like the Earth and the other planets, by accreting the dust and tiny planetoids that formed the primordial nebula around the Sun, some 4.5 billion years ago. Once sufficiently massive, Jupiter's nucleus – like those of the other three giant planets – was then able to attract a large proportion of the hydrogen and helium in the primordial nebula. But light gases such as hydrogen and helium escaped from the Earth's relatively weak gravitational field, whereas Jupiter was able to retain them. More than twice as massive as all the other planets in the Solar System combined, Jupiter exerts enormous gravitational

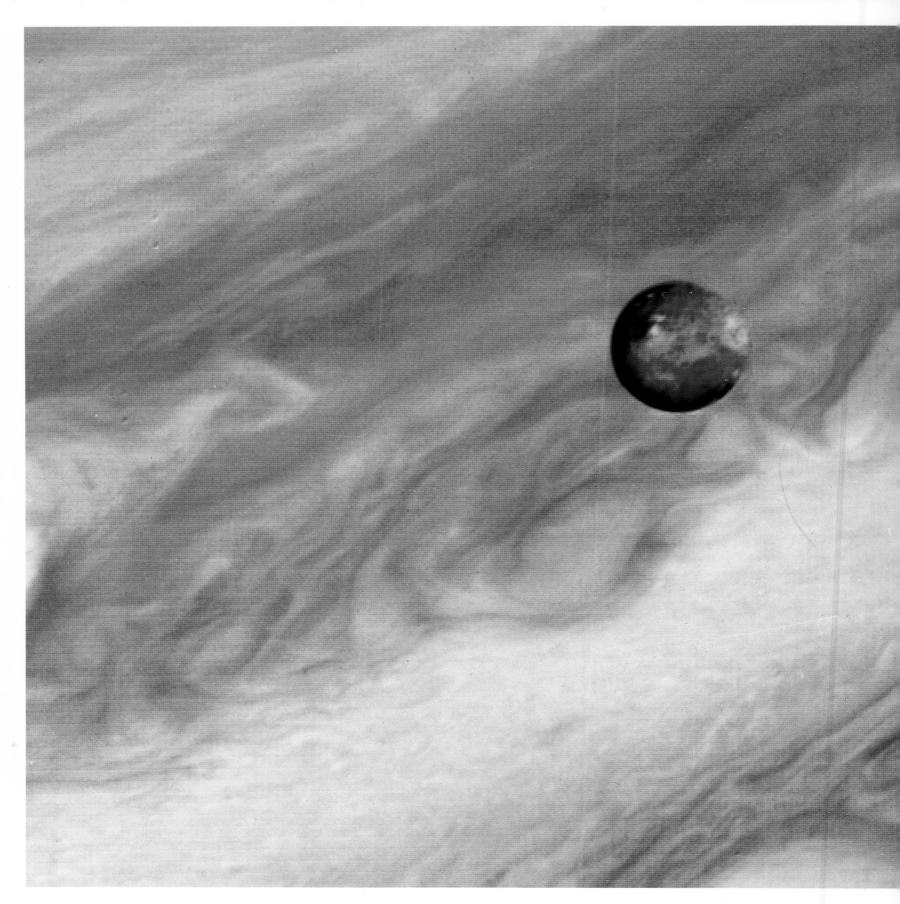

influence over the whole system. It was undoubtedly its gravitational perturbations that prevented a planet from forming where we now find the asteroid belt. Similarly, when the Solar System formed, Jupiter appropriated some of the primordial nebula, which condensed around it to form a complex, rich, satellite system; a sort of miniature Solar System. Sixteen satellites orbit Jupiter (see Appendices). The smallest, the size of Phobos, Deimos, and Gaspra, are probably no more than captured asteroids. The four large Galilean satellites, discovered by the great Italian scholar in 1610, are planets in their own right, because their sizes are comparable with those of the Moon, Mercury and Mars.

Jupiter is much larger than all the other planets; it has a

■ Io SLOWLY CROSSES THE JOVIAN SKY, AND SEEMS ABOUT TO FALL INTO THE VAST OCEAN OF GAS. TOO CLOSE TO THE GIANT PLANET, THE INTERIOR OF THE SATELLITE IS LITERALLY KNEADED BY THE INTENSE GRAVITATIONAL TIDES PRODUCED BY JUPITER. THE

miniature planetary system of its own; and it emits more energy than it receives from the Sun. Should Jupiter be considered a star? Astronomers have wondered about this question for a long time. However, the giant planet does not meet the sole astrophysical criterion that defines a star: the initiation of thermonuclear reactions in the body's core. For this to apply, Jupiter's core would have to reach a temperature of about 10 million degrees Celsius, with a pressure of several billion atmospheres. It is far from this state, because its temperature probably does not exceed 30 000 °C and its pressure 100 million bars. For these thermonuclear reactions to occur in the centre of the planet,

INTERIOR OF IO IS CERTAINLY PARTIALLY MOLTEN.
ITS HIGHLY COLOURED SURFACE IS REPEATEDLY
COVERED BY THE FLOODS OF LAVA ERUPTED FROM
ITS VOLCANOES.

Jupiter would need to be fifty to one hundred times as massive.

Its own source of energy therefore has to be sought elsewhere. Perhaps with its enormous mass, the planet is continuing to shrink and thus heats its core. None the less, Jupiter emits a lot of infrared radiation. If any hypothetical extraterrestrial astronomers observed the Solar System from some distant star with more powerful telescopes than our own, they would see none of the small grey, blue, yellow or red planets that orbit the Sun, but they would certainly detect Jupiter's warm body. They might perhaps deduce that the Sun has just a single planet.

Shoemaker–Levy:

timetable to collision

■ At 22ʰ30ᴹ on 16 July 1994, the first fragment of Comet Shoemaker–Levy crashed into Jupiter. The extraordinary brilliance of the impact was first recorded in the infrared, at Siding Spring Observatory. All the world's telescopes were pressed into service for this unique event in the history of astronomy.

■ SEEN IN THE INFRARED, THE IMPACT SITES STILL APPEAR BRIGHT, AND HENCE HOT, SEVERAL DAYS AFTER THE COLLISIONS. IN THIS IMAGE, FIVE IMPACT SITES ARE VISIBLE IN JUPITER'S SOUTH TROPICAL BELT.

In the absolute blackness of space at the very confines of the Solar System, in that sort of limbo where the Sun is no more than a single star among the other stars, a chance event once disturbed the smooth working of the great cosmic clock. Was it perhaps the passage of a star, at a distance of several light-years from our planetary system, that created slight ripples in the delicate fabric of space-time? Somewhere, some light-months from the Sun, far beyond the last planet, a tiny comet was dislodged from its orbit and, slowly, began to fall towards the centre of the Solar System.

How many hundreds of centuries does such a distant body take to fall into the centre? After a long, accelerating passage, some thousands of billions of kilometres long, the anonymous comet came within the Sun's field of gravitational attraction. Racing towards it at 50 km/s, it brushed past the massive planet Jupiter, and its orbit was affected by the powerful gravitational field of the largest of the planets. Trapped by Jupiter, the comet went into orbit around it. No one knows when this event took place. The comet was too small and too distant to be observed with terrestrial telescopes. It may well have been there, in suspense, for several centuries, or for just a few months.

On the morning of 25 March 1993, the comet suddenly entered human history. In the photographic laboratory of the famous Mount Palomar Observatory, in California, the American astronomers David Levy and Eugene and Carolyn Shoemaker were examining a photographic plate exposed the previous night, during systematic monitoring of the sky. They spotted an extremely strange luminous streak, greatly elongated, and with an exceptionally unusual appearance. Was it the reflection of a bright star, caused by the telescope's optics, or a fault on the emulsion on the plate? That evening, at Kitt Peak in Arizona, another telescope slowly turned to the same region of the sky, just in case ... But the strange body was still there, and the astronomers then realized that this was a new comet – straightaway known as Shoemaker–Levy from the names of its discoverers – that showed an extraordinary behaviour. Its nucleus appeared to be broken into some twenty fragments, arranged in a line, and surrounded by a vast tail of gas and dust. All the observatories in the world were galvanized into action. It is extremely rare to be able to observe a comet in the process of disintegration, and especially to have the chance to do so with the highly sophisticated

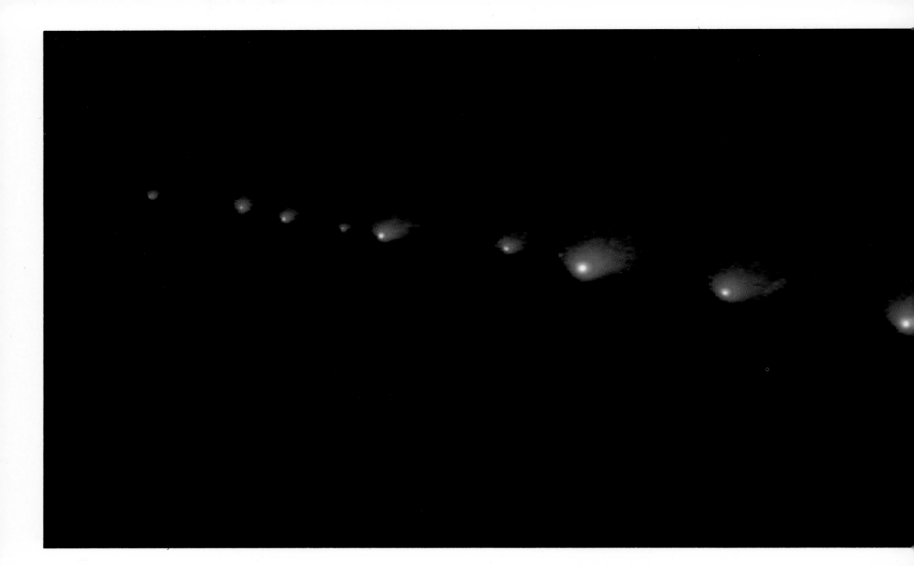

■ Photographed by the Hubble Space Telescope on 17 May 1994, Comet Shoemaker–Levy appeared to consist of about twenty fragments. Scientists do not know the exact size and mass of these fragments. It is

methods available at the end of the 20th century, such as the 10-m Keck Telescope on Hawaii or the Hubble Space Telescope.

But there were more surprises in store for the astronomers. They quickly realized that Comet Shoemaker–Levy was not orbiting the Sun, but Jupiter. The orbit proved to be extremely eccentric. The comet had been discovered near its maximum distance from Jupiter, which was about 50 million kilometres. By measuring the motion of Shoemaker–Levy very accurately, the specialists managed to partially reconstruct its previous path, and calculated that its closest passage to the planet, at a distance of less than 40 000 km, occurred on 8 July 1992, that is, only nine months before its discovery. At that time, the comet was probably intact, and must have resembled an immense oval iceberg, some 5 km in diameter, with a mass of just a few billion tonnes. Moving very fast, the comet passed inside the Roche Limit, the danger zone that surrounds every planet, and where differential gravitational forces cause enormously high strains in the interior of any large-sized body passing through it. The comet disintegrated into twenty-one fragments, each measuring between one kilometre and several hundred metres across, and each of which weighed several hundred million tonnes.

After six months of intensive astrometric measurements, astronomers had to accept the evidence that the path of the comet was going to directly intercept that of Jupiter. They were observing the very last orbit of Shoemaker–Levy. The tiny comet was about to crash into the giant planet. To these scientists, this was an event of historical significance, and astronomers did not fail to appreciate the luck that had given them the chance to observe it directly. This was, in fact, the first time that they would witness the collision of two worlds. Never before, since the invention of the astronomical telescope by Galileo at the beginning of the 17th century, had an observer been able to see such an event. Over a longer time-scale, however, such as a hundred million years, collisions of comets and asteroids with planets are very frequent. Our whole planetary system has been slowly built by impacts, and the number of craters found on Mercury, the Moon, Mars, and the satellites of the giant planets number in the billions.

On Earth, it was a question of wait and see. What would happen? Nothing, said some planetologists. To them, the comet, with not one trillionth of the mass of Jupiter, would simply be vaporized when it hit the planet's turbulent atmosphere. What effect would the flapping of a seagull's wings have in the middle of a storm on the atmospheric conditions on Earth? Other researchers, in contrast, estimated that a body, however small, encountering a planet at a velocity of 216 000 km/h, ought to upset its thermal and dynamical equilibrium and leave traces …

In the spring of 1994, observations of Shoemaker–Levy intensified. Astronomers continuously followed the train of fragments, which continued to spread out and to break up. At the time of its discovery, Shoemaker–Levy looked like a beautiful set of pearls, some 160 000 km long, but by the beginning of July 1994 the twenty-one fragments on this suicide course had spread out to more than five million

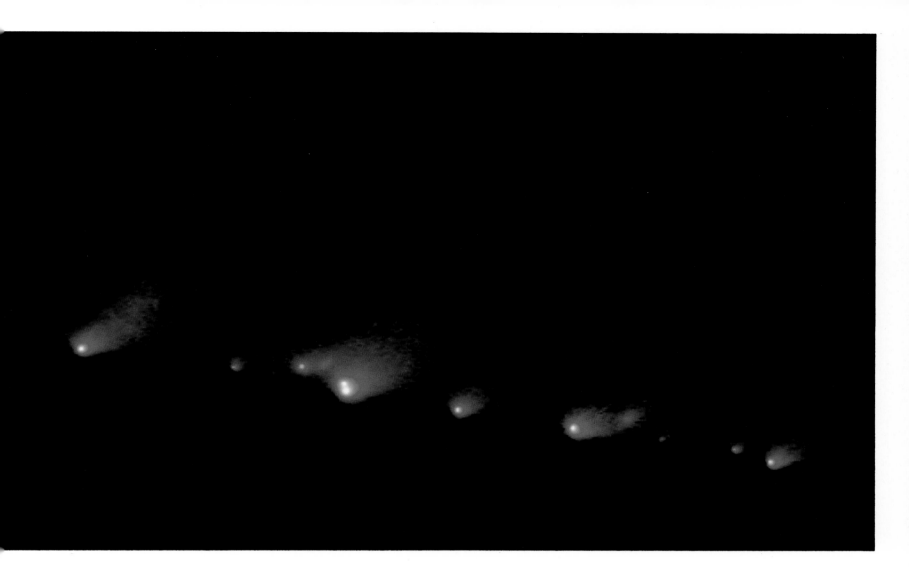

ESTIMATED, HOWEVER, THAT THEIR DIAMETERS LIE BETWEEN SEVERAL HUNDRED METRES AND ONE KILOMETRE. IN THIS IMAGE, THE COMET EXTENDS OVER A DISTANCE OF MORE THAN ONE MILLION KILOMETRES.

kilometres. In fact, rather than a single collision, astronomers expected a bombardment that would last a week, from the 16th to 20th July. Unfortunately, all the impacts would take place just on the far side of the boundary between the visible and invisible sides of Jupiter, and only their effects would be detectable ten minutes later, when the impact sites, illuminated by the Sun, would appear around the limb of the planet as it rotated. Theorists had made these predictions in celestial mechanics as a matter of urgency six months earlier, but were to be proved correct to within a few minutes.

Despite the Cassandras, who continued to maintain that the collision would not produce any effects visible from Earth, more than 700 million kilometres from Jupiter, observers readied their telescopes. The scientific forces deployed were formidable, and unique in the history of astronomy. On Earth, all the largest telescopes were requisitioned for the event. In space, both the Hubble Space

■ ON 20 JULY 1994, JUPITER'S SOUTHERN HEMISPHERE APPEARED HIGHLY DISTURBED BY THE COLLISIONS. THIS INFRARED IMAGE WAS TAKEN AT A WAVELENGTH OF 2.5 μM WITH THE 3.5-M DIAMETER TELESCOPE AT CALAR ALTO IN ANDALUSIA, AND TRANSMITTED IMMEDIATELY OVER THE INTERNET TO ALL THE OBSERVATORIES IN THE WORLD.

Telescope and the Galileo spaceprobe were programmed to follow the collisions.

On 16 July, every telescope was aimed at Jupiter, although the fragments of the comet were utterly invisible, having been lost for several days in the blinding light from the planet. Suddenly, at 20^h30^m UT, an infrared flash of extraordinary brilliance lit up the limb of the planet. The first fragment of Comet Shoemaker–Levy had smashed into Jupiter and had literally set its atmosphere alight. It was a cataclysm of unheard of violence, and of a totally unexpected magnitude.

The lump of material, consisting of ices and interplanetary dust, encountered the upper layers of Jupiter's atmosphere at 60 km/s and disintegrated within a few seconds, reaching a depth of about one hundred kilometres, corresponding to a pressure level of about 1 bar. In the wake of the impact, a plasma at nearly 10 000 °C, hotter than the surface of the Sun, rose into the atmosphere and formed a monstrous mushroom cloud,

■ THIS IS WHAT ONE COLLISION SITE LOOKED LIKE. BECAUSE THE PLANET HAS ROTATED, THE IMPACT, AT FIRST VISIBLE IN INFRARED ON THE PLANET'S LIMB, IS NOW VISIBLE WITH OPTICAL TELESCOPES. HERE, ON 18 JULY 1994, ONE HOUR AFTER IMPACT, THE HUBBLE SPACE TELESCOPE HAS CAPTURED THE ENORMOUS MUSHROOM CLOUD THAT FORMED ABOVE THE SEVENTH IMPACT, THE MOST VIOLENT OF ALL.

more than 3000 km high, whilst a colossal shockwave raced outwards at 15 000 km/h, raising a wave of gas that swept all before it.

Ten minutes later, optical telescopes were able to see a vast, extremely dark, expanding area in Jupiter's atmosphere. At the end of an hour, the hot cloud had reached its maximum extent. It was large enough to cover the whole Earth. The energy released by the impact was several million megatonnes! Watching in amazement, astronomers in Australia, Spain, France, Chile, Antarctica, Hawaii, and elsewhere, recorded a film of the

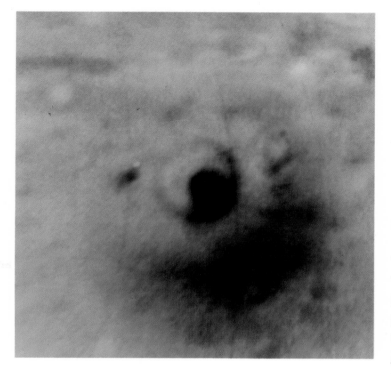

■ A CLOSE-UP PHOTOGRAPH FROM THE SPACE TELESCOPE OF THE CLOUDS CREATED 90 MINUTES AFTER THE SEVENTH IMPACT, ON 18 JULY 1994. THIS GIGANTIC, RAPIDLY EXPANDING, GASEOUS MUSHROOM CLOUD WAS THEN 3000 KM HIGH AND 12 000 KM IN DIAMETER. THE ENERGY RELEASED BY THE IMPACT WAS SEVERAL MILLION MEGATONNES.

collision, passing the baton to one another via the Internet's computers as Jupiter set below their local horizons. Over subsequent days, the other fragments fell, one by one, into Jupiter's atmosphere, leaving impact scars of greater or lesser size, which sometimes overlapped, and which were slowly distorted by jovian winds. The size of the largest 'hole' left in Jupiter's atmosphere, that of the seventh impact, exceeded 30 000 km. Despite the violence of winds on Jupiter, and the power of the currents in the atmosphere of a planet 318 times as massive as the Earth, and which is rotating more than twice as fast, atmospheric mixing did not rapidly erase the dark traces left by the comet. By spring 1995, these were still visible, encircling the planet as a dark cloudy band within the South Tropical Zone.

The Galileo spacecraft, which was approaching Jupiter at the time of the collisions, arrived in time, at the very beginning of 1996, to study the dark clouds produced by the impacts. According to some experts, the planet would not resume its normal appearance for several years.

When the fragments of Comet Shoemaker–Levy im-

pacted Jupiter, they behaved as the Galileo atmospheric probe would behave when it was released to plunge into the atmosphere to study it in depth. The millions of electronic images and the spectrograms obtained over a whole range of wavelengths during that historic week of observations have still to be analysed in detail. But quite apart from progress in our knowledge of the chemical composition of Jupiter's clouds; of the origin of comets; of the stability of their orbits; of the size, density, composition, and mechanical strength of cometary nuclei; and of their fragmentation, it is perhaps a re-evaluation of the statistics relating to cometary and asteroidal impacts on the planets in the Solar System that will most interest the experts. As well as the effects that the collision with a comet like Shoemaker–Levy would have – on Earth.

■ ON 22 JULY, AFTER THE END OF THE BOMBARDMENT, THE HUBBLE SPACE TELESCOPE PHOTOGRAPHED THIS SERIES OF IMPACTS, LYING SOUTH OF JUPITER'S GREAT RED SPOT. AT BOTTOM RIGHT OF THE IMAGE, THE SEVENTH IMPACT SITE IS DISAPPEARING AROUND THE LIMB OF THE PLANET. SEVERAL WEEKS LATER, THE DARK CLOUDS FORMED DURING THE COLLISIONS HAD SPREAD RIGHT ROUND THE SOUTH TROPICAL BELT.

Io:

the volcano planet

■ THE CLOSEST OF THE GALILEAN SATELLITES IS JUST
SLIGHTLY LARGER THAN THE MOON. BUT IO, ORBITING
VERY CLOSE TO THE LARGEST PLANET IN THE SOLAR
SYSTEM, IS CONTINUALLY SUBJECT TO ITS GRAVITATIONAL
EFFECTS. ATTRACTED ON ONE SIDE BY JUPITER, AND ON
THE OTHER BY JUPITER'S THREE REMAINING MAJOR
SATELLITES, IO IS DEFORMED, KNEADED AND
HEATED BY THE ABUNDANT GRAVITATIONAL
ENERGY. THIS IS RELEASED FROM THE
SATELLITE BY INCESSANT VOLCANIC
ERUPTIONS.

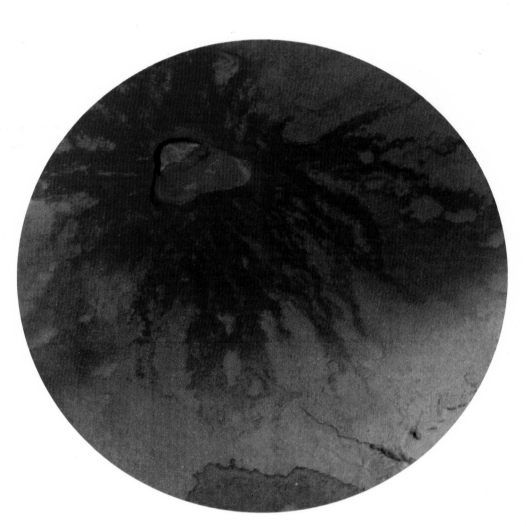

■ MAASAW PATERA, ONE OF IO'S GREAT VOLCANOES, NOT FAR FROM THE SOUTH POLE. ITS MORPHOLOGY RESEMBLES THAT OF THE GREAT HAWAIIAN SHIELD VOLCANOES. BUT THE SCALE IS COMPLETELY DIFFERENT: ITS SUMMIT CALDERA MEASURE 50 KM IN DIAMETER.

The yellow ground shakes and feels as if it is about to give way. In a dark sky, a gigantic, multicoloured, oval balloon, striped with parallel bands of cloud, floats unmoving, tilted at an angle. In absolute silence – there is no atmosphere here – a geyser of molten sulphur erupts above a lake of black lava. The column of gas seems to be attacking the sky. The fountain rises for nearly ten minutes, then, at a height of about 30 km, the fiery spray slowly falls back in a splendid plume, which eventually hits the ground. Io is the volcano planet. At this precise instant, as you read these lines, its great volcanoes, Prometheus, Maasaw Patera, Pele, Loki, or Ra Patera are sending jets of lava at speeds of more than 3000 km/h into the sky.

We are on a strange world, witnessing a spectacle worthy of science fiction, which unfolds (without the slightest sound) scenes of unbelievable beauty and violence. Silhouettes of volcanoes appear along the horizon, and the largest of these is emitting a halo of gas and dust in front of the stars. Towards the east, a vague, bluish cloud rises from a lava lake and hides the stars for a moment before fading away. It is cold. In the shadow of the great volcanoes, sulphur dioxide has been deposited on the ground, forming great striking swirls of frost. It is early afternoon on the small planet Io, and the temperature is close to −150 °C. The tawny-coloured plains that surround us, patterned like leopard-skin, are stratified, striped with yellow, orange, and red sediments, which lie on top of one another and overlap with no apparent logical explanation.

Io is the closest of Jupiter's four Galilean satellites to the planet. At a distance of 421 000 km, it takes less than two days to orbit the giant body, which appears forty times as large in its sky as the Moon does when seen from Earth. Io has physical characteristics that are comparable with those of the Moon. First, its mass: 89 billion billion tonnes, as against 73 billion billion tonnes for the Moon. Second, its density: 3.57 as against 3.34. Finally, its diameter: 3640 km for Io, while that of the Moon is 3476 km.

Nevertheless, the comparison between the two small worlds stops there. Although the surface of the Moon is covered in craters, there is, in contrast, not the slightest trace of any impact on Io! Yet impact craters bear witness to the Solar System's origins, when the plane of the disk of dust and gas, from which the Sun and its retinue of planets formed

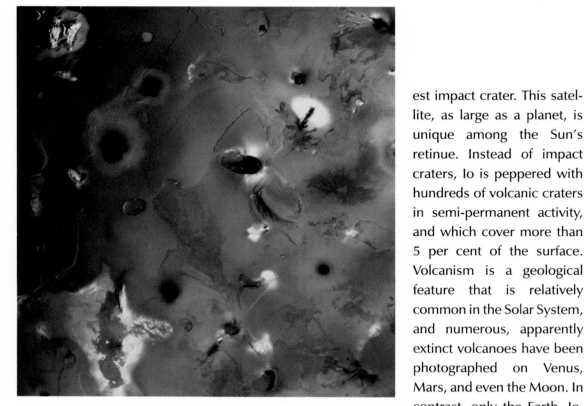

■ Io is surrounded by a thin atmosphere, whose pressure is uneven and increases above the major volcanic centres. This extremely tenuous atmospheric halo, permanently fed by gas from the volcanic eruptions, is never more than one millionth of a bar. Io's atmosphere probably has the lowest density of that found on any of the bodies in the Solar System.

simultaneously, was criss-crossed by planetoids, travelling in all directions, crashing into one another, sometimes breaking up, sometimes sticking together, and finally falling one by one onto the largest of the bodies. This meteoritic bombardment was at its height during the first half-billion years of the Solar System's existence. Craters are, in fact, the commonest geological element among the planets, and are the best tool that planetologists have. Their number, size, depth, shape, ejecta, erosion, and the way they overlap, allow researchers to literally travel back in time, to the planets' very distant past. With the exception, of course, of the four giant planets, which have no proper surfaces to speak of, and Titan, which is hidden by a thick atmosphere so its surface remains unknown, all the planets, all the satellites, and all the asteroids known in the Solar System are pock-marked by impact craters. The smallest worlds and those with least geological activity, such as Mercury and the Moon, are those that have the most craters. On the larger terrestrial planets, endowed with atmospheres, such as Venus, Earth, and Mars, impact craters are destroyed with time. On Earth, very few traces of these ancient impacts have been left, thanks to the action of water, frost, aeolian erosion, sedimentary layers deposited by billions of generations of living creatures, basalt flows erupted from volcanoes and, above all, through the slow movement of continental plates. Only the most recent impacts have left their mark on the Earth's surface. On Io, however, there really is not the slight-est impact crater. This satellite, as large as a planet, is unique among the Sun's retinue. Instead of impact craters, Io is peppered with hundreds of volcanic craters in semi-permanent activity, and which cover more than 5 per cent of the surface. Volcanism is a geological feature that is relatively common in the Solar System, and numerous, apparently extinct volcanoes have been photographed on Venus, Mars, and even the Moon. In contrast, only the Earth, Io, and to a lesser extent, Triton exhibit active volcanism today. That on Io is utterly rampant. How can such a small body – which is twelfth in order of size of the bodies in the Solar System – produce so much energy? The answer to this question lies in the particular location of Io in Jupiter's system. It is very close to the giant planet, which is twenty thousand times its mass. Intuitively, it does not take much to imagine that it must be subjected to intense gravitational tides caused by Jupiter. Yet, because the orbit of Io is almost perfectly circular, these tidal forces should not, basically, have any affect. Researchers have found, however, that two of Jupiter's other large satellites, Europa and Ganymede, act to cause perturbations by slightly modifying the orbit of Io around Jupiter as they complete their own orbits around the planet. Io is thus kneaded by tides, whose amplitude is unique in the Solar System, becomes deformed and heats up. The interior of the satellite must be an actual ocean of magma, on which floats the solid crust, which consists of silicate rocks. These seas of

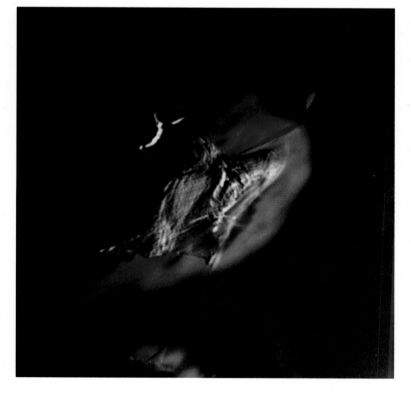

■ Haemus Mons, which lies not far from Io's south pole, is one of the largest mountains in the Solar System. Completely covered in sulphur frost, it is seen here just as it emerges from night-time darkness. Measuring nearly 200 km in diameter at its base, Haemus Mons, is probably more than 10 000 m high. Its geological origin is unknown.

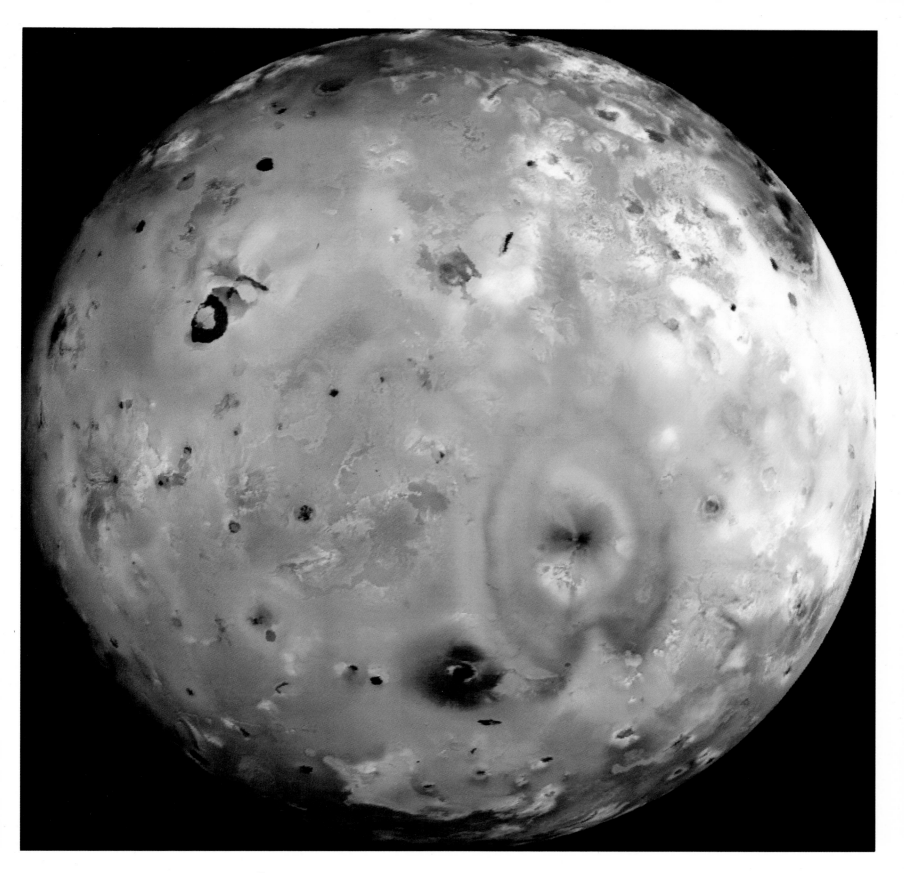

lava, which are slowly stirred by the tides, break through to the surface here and there, and spread out over the yellow- and orange-coloured plains.

The great volcanoes on Io have little in common with those on Earth. We still do not know their heights – which probably lie between 1000 and 5000 m – but their diameters are well known. On Io, there are more than two hundred calderas that are more than 20 km in diameter. By comparison, the Earth, which is the most volcanically active planet after Io, and which has a surface area that is more than three hundred times as great, has no more than about fifteen volcanic massifs of this size. The largest volcanoes on Io are monstrous. Ra Patera, for example, measures 250 km in diameter, and the flows of very fluid lava from it have dug dark channels in the flanks of the volcano and extend for distances of more than 200 km. The volcano Loki is even more active and strange. Beneath its summit, an almost circular lava lake, measuring 250 km in diameter, is permanently fed with molten sulphur rising from the depths. A sort of 'ice floe' consisting of solid sulphur floats on this lake, and occasionally 'icebergs' become detached and finally melt into the lava, which is at a temperature of 200 °C. Loki, like all the other volcanoes on Io, was discovered by the Voyager spaceprobes between March and July 1979. Between the two encounters, the volcanic activity changed considerably. Some volcanoes had become extinct, others had revived. Since then, astronomers have been able to observe several volcanoes from a distance. Only

PELE IS THE MOST POWERFUL VOLCANO KNOWN ON IO. OBSERVED FOR SEVERAL CONSECUTIVE WEEKS BY THE VOYAGER 1 SPACEPROBE, IT DISPLAYED PLUMES OF GAS AND DUST THAT REACHED SOME 300 KM IN HEIGHT AT SPEEDS OF 1800 KM/H. MORE THAN ONE MILLION SQUARE KILOMETRES OF THE REGION AROUND PELE ARE COVERED IN VOLCANIC DEPOSITS.

■ DEPENDING ON WHETHER IO IS IN NIGHT-TIME DARKNESS, HEATED BY THE SUN, OR IS SUBJECT TO A VOLCANIC ERUPTION, THE SULPHUR THAT COVERS IT TAKES ON TINTS THAT ARE CREAMY, TAWNY, OR ALMOST BLACK. THE AVERAGE TEMPERATURE OF THE PLANET IS AROUND −150 °C. AT THE BOTTOM RIGHT OF THIS IMAGE, THE VOLCANO LOKI IS IN FULL ERUPTION AND IS SURROUNDED BY A HALO OF BRIGHT VOLCANIC DEPOSITS.

Loki has remained in permanent eruption over all that time!

When compared with terrestrial criteria, eruptions on Io seem like something out of Dante. The most violent, like those of Prometheus, send vast gaseous plumes more than 300 km into space, which then fall back to the surface, forming rings of molten sulphur 1000 to 2000 km in diameter! Each of the most powerful eruptions of Prometheus rips nearly 10 000 t of molten rock from Io every second. Every year, a not inconsiderable proportion of the mass of the body – perhaps several hundred billion tonnes – is expelled from the depths, before falling back and coating the surface. On average, more than one tenth of a millimetre of volcanic material is deposited every year on Io. Over a long period of time, Io's plains have been completely covered by sulphur, which is the volcanoes' principal material. These lava flows and fine rain of sulphur have eventually blotted out the whole of the planet's past. Because, at the current rate, every million years tens of metres of coloured strata have covered the planet in a tawny-coloured sulphurous blanket. Geologically, the surface of Io is the youngest in the whole Solar System, and kilometres of sulphur now cover the ancient impact craters that the experts searched for so long in vain. Despite their gigantic size, and their highly impressive ability to remodel and reconstruct their surroundings, the power of the volcanoes on this small body is relatively low. As on the Moon, gravity here is about one sixth of that on Earth. In addition, the more-or-less total absence of any atmosphere helps with the ejection of volcanic material high into the sky. Placed on Earth, the most powerful volcanoes on Io would pale into insignificance alongside monsters such as Krakatoa or Tambora, which terrified earthlings during the 19th century or, more recently, Pinatubo, which polluted the whole of the Earth's atmosphere in just a few months with its wind-borne volcanic dust.

Ever since the visits by the Voyager spaceprobes in 1979, astronomers ceaselessly tried to observe Io's volcanoes from Earth. More than sixteen years were to pass before another robot returned to Jupiter's system. It was not until the end of 1995 that the Galileo spacecraft was captured by Jupiter's powerful gravitational field. In the meantime, however, new methods of observation had been developed on Earth.

First, from 1990 onwards, the Hubble Space Telescope, from its orbit around the Earth at an altitude of 600 km, was able to follow volcanic eruptions on Io with, at times, an image quality that was close to that obtained by the Voyagers! Subsequently, certain observatories have succeeded in doing the same from ground level. Incredible as it may seem, some planetologists have specialized in studying the volcanoes on this other body, which they examine, month by month, from a distance of more than 500 million kilometres. To detect these much talked-about volcanoes, French and Canadian astronomers have develop-

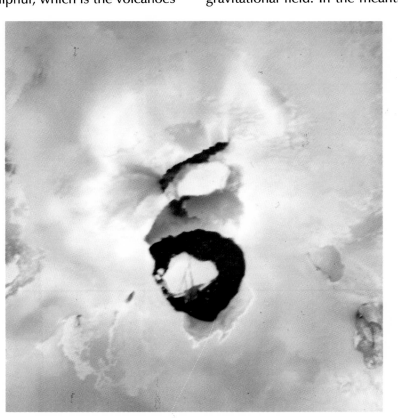

■ THE PHOTOGRAPH TAKEN BY THE VOYAGER 1 SPACEPROBE IS PROBABLY ONE OF THE MOST EXOTIC IMAGES EVER OBTAINED FROM SPACE. IT SHOWS THE LARGE MOLTEN SULPHUR LAKE INSIDE LOKI. VERY DARK, IT MEASURES NEARLY 250 KM FROM ONE SHORE TO THE OTHER. FLOATING IN THE LAKE, AN ICEBERG, PROBABLY CONSISTING OF SOLID SULPHUR, IS SLOWLY BREAKING UP IN THE HOT LIQUID.

ed a new, surprising, and remarkably elegant method. Their 3.6-metre telescope, installed at the summit of the volcano Mauna Kea, in the Hawaiian islands, is fitted with a system of adaptive optics, which consists of a flexible mirror, linked to an ultrafast computer. The adaptive-optics system follows the movement of the atmosphere and corrects it fully, by deforming the mirror to match the turbulence, so that the telescope always gives an image of perfect quality, as if it were installed in the vacuum of space. To increase their chances, the astronomers observe Io with an infrared camera, which is particularly sensitive to the heat emitted by the volcanoes. Finally, to increase the quality of the scientific data obtained still further, the observations are made when Io passes into Jupiter's shadow. These eclipses plunge Jupiter's satellite into darkness, which means that the active volcanoes are all the more prominent on the images obtained by the telescope.

At the end of the 20th century, working together, the Galileo probe, the Hubble Space Telescope, and ground-based telescopes have allowed us to improve our knowledge of the volcano planet. For example, the strength of the tides that continually knead the centre of Io has been determined as 10 000 billion watts, an amount of energy – twice that produced by the Earth – which appears colossal when one thinks that our blue planet is more than sixty times as massive as Jupiter's satellite! In another surprising discovery, it has been determined that some lava flows on Io have a temperature of 1500 °C, which sug-

■ THESE IMAGES OF IO FORM AN EXTRAORDINARY RECORD. THEY WERE OBTAINED BY VOYAGER 1 IN 1979 *(TOP)*, AND BY GALILEO, IN 1996. OVER A PERIOD OF SEVENTEEN YEARS SUBTLE DIFFERENCES IN THE APPEARANCE OF THE SURFACE ARE VISIBLE: WHOLE REGIONS OF THE SATELLITE HAVE BEEN ALTERED COMPLETELY.

gests that lava on Io does not consist solely of sulphur. Only molten silicates that originate at depth in the satellite can actually attain such temperatures, which is a Solar-System record. Over a period of five years, Galileo discovered and observed nearly one hundred hot spots. In certain cases, the volcanic eruptions could be followed for several months at a stretch. The most important of these was first detected by the Space Telescope. In June 1997, a splendid plume of sulphur was detected, rising some 120 km above the surface. The volcano Pillan Patera had reawakened. The scientists then aimed Galileo's cameras at the large volcano, and compared the images with those that the space-probe had obtained before the eruption. The surroundings had been completely changed by the eruption, because from then on a black spot, 400 km in diameter and probably a lake of molten sulphur, surrounded Pillan Patera.

At the end of 1999, despite the great risk that it would give up the ghost, being overcome by Jupiter's radiation belt, the Galileo spaceprobe made two close fly-bys of Io, to observe its surface in detail. The first fly-past, despite several failures and technical hitches caused, once again, by Jupiter's radiation belt, took place perfectly at a distance of just 610 km, on 10 October 1999. One of the several tens of images obtained, with a resolution of about 10 m, showed a lava field – with an astonishing resemblance to a terrestrial lava flow – that had recently been erupted from Pillan Patera, which had been in semi-permanent eruption since June 1997. A second

■ SINCE THE END OF 1995, GALILEO HAVE BEEN PERMANENTLY MONITORING THE VOLCANIC ACTIVITY ON IO. ON THIS IMAGE, TAKEN IN JUNE 1997, TWO ERUPTIONS ARE VISIBLE. THE PLUME FROM PILLAN PATERA APPEARS ON THE LIMB, WHILE PROMETHEUS PATERA IS VISIBLE ON IO'S DISK, AT THE CENTRE OF THE IMAGE.

fly-past of Io occurred on 25 November 1999. The Galileo probe, flying just 300 km above the surface, passed through the plume of the Pillan volcano and recorded new data as it did so. By comparing the images taken of the same region of Io at an interval of just 45 days, the planetologists were able to assess the violence of volcanic eruption on the tiny body. Certain relief features had disappeared, buried beneath vast lava flows. Finally, the height of the largest active volcano in the Solar System could be determined. Prometheus Patera measures more than 16 000 m. After these two fly-pasts, which could well have been fatal, the probe still continued to function. NASA's engineers and scientists decided to continue the Galileo mission until the end of 2000. In December of that year, the joint American and European Cassini-Huygens space-probe had a Jupiter encounter. This was an ideal occasion to observe the surface of Jupiter and its magnetic field simultaneously with two different space-probes. The astronomers' objective was to produce images and three-dimensional charts of the complex jovian system from all the data that was collected.

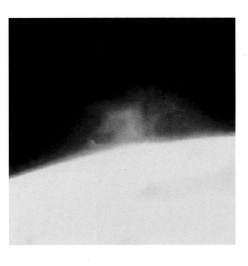

■ A CLOSE-UP OF PILLAN PATERA. THE VOLCANIC ERUPTION, WHICH STARTED IN JUNE 1997, ERUPTED MOLTEN SULPHUR TO A HEIGHT OF MORE THAN 120 KM FOR SEVERAL MONTHS.

Io's continual eruptions eject billions of tonnes of lava, dust and gas into space. Some of this does not fall back immediately, and forms an extremely tenuous, and highly variable, atmosphere around Io. The atmospheric pressure – if one can speak of it in those terms – is between one millionth and one billionth of a bar! Nowhere else in the Solar System is there so evanescent an atmosphere, except perhaps on Triton or Pluto. This 'micro-atmosphere' is also affected by Jupiter's immense magnetic field. Its lines of force permanently flow through the smaller body and tend to capture atoms of sodium and sulphur from its surface. This is the tribute that the strangest body in the Solar System pays to its powerful and highly influential neighbour.

Europa, Ganymede, Callisto:

terrae incognitae to explore

■ THE PLANET JUPITER IS ACCOMPANIED BY A RETINUE OF FOUR GIANT SATELLITES: IO, EUROPA, GANYMEDE, AND CALLISTO, ALL OF A SIZE AND MASS COMPARABLE WITH THOSE OF THE MOON. EUROPA IS THE SMALLEST AND THE LEAST MASSIVE OF THESE FOUR BODIES. ITS EXTRAORDINARILY SMOOTH SURFACE, ALMOST COMPLETELY LACKING IN CRATERS, IS VERY WHITE AND REFLECTIVE AND UNDOUBTEDLY CONSISTS OF ICE. EUROPA ITSELF IS DEVOID OF ANY ATMOSPHERE.

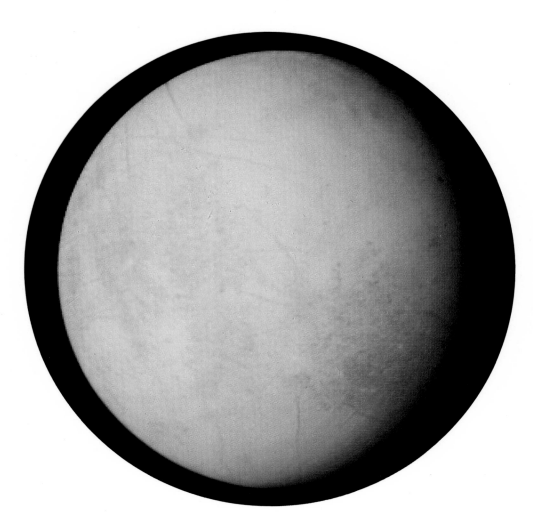

■ SEEN FROM A DISTANCE, THE SMALL BODY OF EUROPA APPEARS AS SMOOTH AS A BILLIARD BALL. NEVERTHELESS, ANALYSIS SHOWS THAT THE SURFACE OF THE BODY HAS COMPLEX ALBEDO MARKINGS AND EVEN VERY SLIGHT RELIEF.

The landscape is utterly, desolately, flat. No high mountains or volcanoes on the horizon, no canyons, no valleys filled with morning mist, no fields of dunes reddened by the setting Sun. Under a black sky, with not the slightest trace of cloud, everywhere the plains of Europa look the same: a desert of bright ice, as smooth as a mirror. A thousand kilometres farther on, towards the south, the west, or the east, the panorama remains the same; occasionally enlivened, from time to time, by a small crater left by a meteorite impact, 1000 or 10 000 years ago. No other body in the Solar System exhibits such a monotonous surface. The whole surface of Europa is like an immense ice sheet, extending for tens of millions of square kilometres, so pale and smooth that it would make your head swim.

The sights, on Europa, are overhead, in the airless sky. Jupiter, just under 700 000 km away, slowly turns on its axis, showing off its violent, silent storms, capable in their lifetimes of swallowing Mars, the Moon, and the Earth (and all its theories), at a single gulp. Not far from the enormous disk of Jupiter – seen from here, the giant planet is twenty times as large as the Moon as seen from Earth – the brilliant disk of Io,

the volcano planet, may be seen. When it passes close to Europa, at a distance of just 250 000 km, Io clearly shows its disturbing volcanoes, which form red or black patches against the yellow or brilliantly white plains of this tiny body, melted by the gravitational tides of its too-powerful neighbours. In Europa's sky, the orange-tinted disk of Io is nearly twice as large and ten times as bright as the Moon when seen from Earth. At night, the volcano planet illuminates the frozen plains with tawny-coloured light and reflections. Without an atmosphere to retain the heat of the distant Sun, the temperature of this icy desert never exceeds − 160 °C.

Europa is the smallest and least massive of Jupiter's four giant satellites. Its diameter is 3140 km, and its mass is over 48 billion billion tonnes, with a density that is almost equal to 3. Revolving at a distance of 671 000 km from Jupiter, on a quasi-circular orbit, which it completes in three-and-a-half days, Europa is second in the sequence of Jupiter's large satellites, coming after Io, and before Ganymede and then Callisto, which is the last of the four Galilean satellites.

Europa resembles nothing so much as an almost perfectly white billiard ball, with an albedo of nearly 70 per cent, which

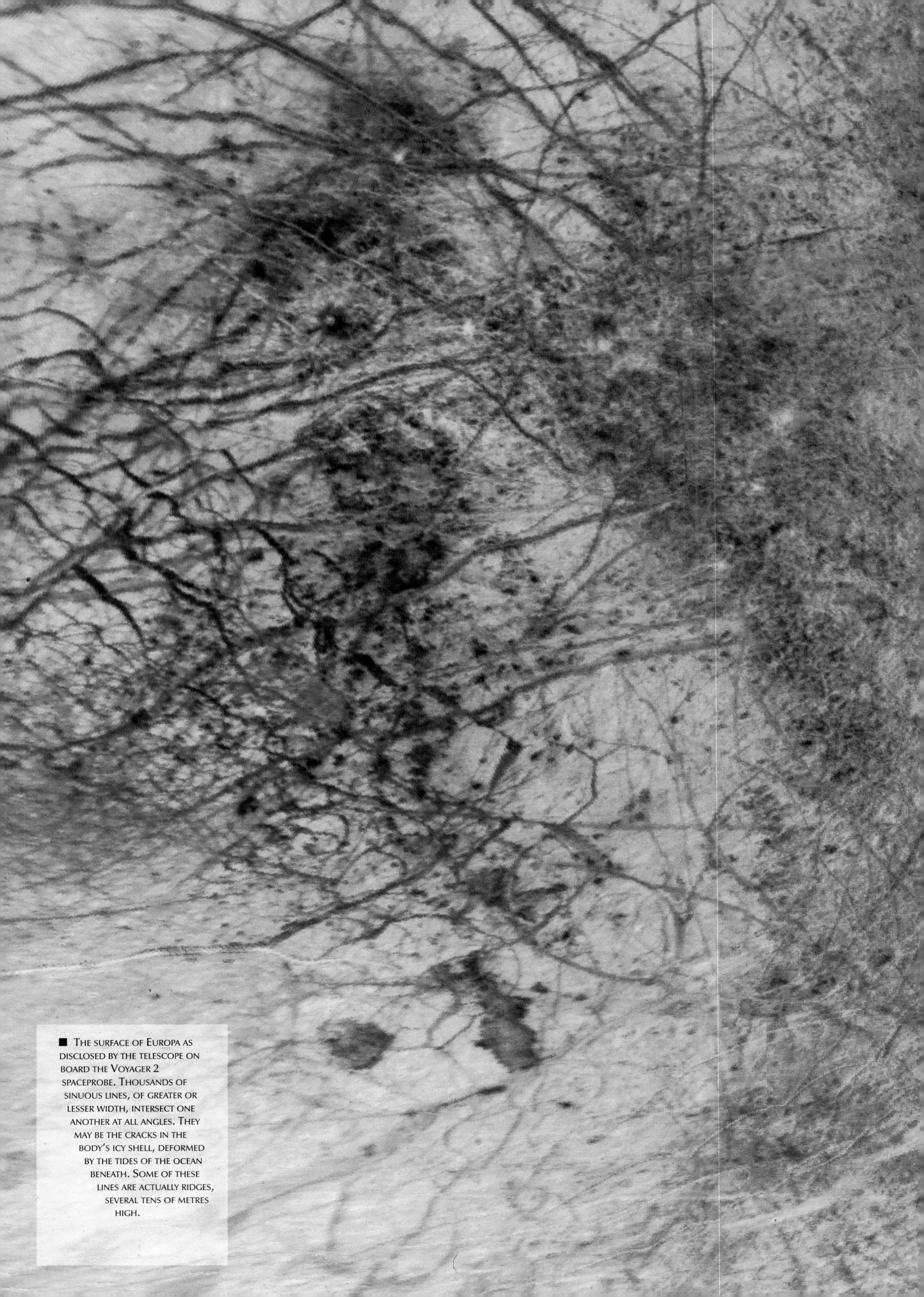

■ THE SURFACE OF EUROPA AS
DISCLOSED BY THE TELESCOPE ON
BOARD THE VOYAGER 2
SPACEPROBE. THOUSANDS OF
SINUOUS LINES, OF GREATER OR
LESSER WIDTH, INTERSECT ONE
ANOTHER AT ALL ANGLES. THEY
MAY BE THE CRACKS IN THE
BODY'S ICY SHELL, DEFORMED
BY THE TIDES OF THE OCEAN
BENEATH. SOME OF THESE
LINES ARE ACTUALLY RIDGES,
SEVERAL TENS OF METRES
HIGH.

means that the surface of the body, essentially consisting of ice, reflects 70 per cent of the light from the Sun. On looking more closely, despite the fact that its surface is nearly perfect – to scale, it is actually smoother than a billiard ball! – Jupiter's second largest satellite appears to be covered in a complex network of fine, sinuous lines and bands, which cover it entirely. These lines and bands, which are clearly visible on the electronic images obtained by the two Voyager probes, are, in reality, very close in colour and brightness to the general surface of the satellite. The difference in albedo between these dark, yellowish marks and that of the overall, ivory colour of the satellite is, in fact, less than 10 per cent. From a distance, explorers approaching Europa would see a perfectly smooth, white sphere. Only close up would they discover the thousands of lines, of various widths, that cover the whole of the surface. The narrowest of these measure just a few kilometres wide, but may stretch, in long geodesic curves following the curvature of the planet, for hundreds of kilometres! Other, wider bands, such as Asterius Linea or Minos Linea, may be more than 3000 km long. Seen from a distance, Europa resembles the shell of an egg that has been cracked into a thousand pieces and then patiently reassembled without a visible break. This puzzling world fascinates planetologists. They have not discovered on its surface a single one of the geological and topographic features that are normally found on a solid body. Like Io, but for completely different reasons, Europa has practically no impact craters. Just three or four large impacts have been detected on its surface; evidence of its extreme youth. The cratering rate on Europa is practically the same as that of the Earth, which is, with the notable exception of Io, the lowest in the whole Solar System. Europa has less than one per cent of the craters per unit surface area found on the Moon, Mars, Ganymede, and Callisto. Like Io, Europa exhibits an extremely young surface. The age of the smooth crust of ice that covers it can hardly be more than a few tens of millions of years. What happened on Europa to turn it into this enigmatic ball, with no apparent trace of its past?

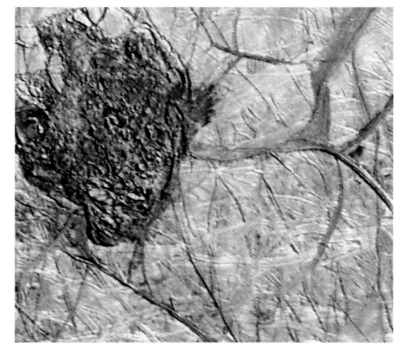

■ THE SURFACE OF THE SATELLITE EUROPA IS, GEOLOGICALLY SPEAKING, VERY RECENT. THE ICE THAT COMPLETELY COVERS THE SATELLITE SHOW PRACTICALLY NO IMPACT CRATERS. THIS IMAGE, TAKEN BY THE GALILEO PROBE IN 1998, REVEALS WHERE MATERIAL, WHOSE NATURE REMAINS UNKNOWN, HAS EMERGED FROM THE INTERIOR OF EUROPA.

Because the density of Europa, which lies between 2.98 and 3.02, has been accurately determined, we know that we are dealing with a body that is essentially rocky. Researchers assume that when it formed around Jupiter, some 4.5 billion years ago, its core, heated by radioactive elements, slowly degassed and released light components, such as water molecules. This water may have derived from the dehydration of the silicates that are the major component of Europa's rock. Around the small world, which was probably surrounded at that period by a thin atmosphere, a real ocean would have formed, heated from the interior, but exposed to the cold of space. Several tens of kilometres deep, it must have been criss-crossed by gigantic waves raised by jovian tides or by the regular impact of asteroids. With time, the atmosphere disappeared and the meteoritic bombardment diminished in intensity, causing this strange extraterrestrial ocean to freeze over. The fractured surface that astronauts may perhaps one day survey is probably the result of the restlessness of this world's primordial ocean, always present beneath the surface, creating pressure and tension within the ice, and perhaps erupting into the sky from time to time as immense geysers, falling as crackling ice onto the ice sheet. In fact, scientists assume that, like the core of Io, that of Europa is distorted, kneaded, and thus heated by the gravitational tides caused by Jupiter, Io, and Ganymede. This internal heating of the small world is probably sufficient to maintain most of its water in a liquid form beneath a surface layer of ice.

Between 1995 and 2000, the main mission of the Galileo probe was the detailed observation of Europa, which it flew past several times, eventually at the record distance of just 200 km! The quality of the images of this icy satellite obtained by the American spaceprobe may truly be described as stupefying. On the closest images, details just 6 to 9 metres across are detectable. And what Galileo discovered on the surface of Europa was just what planetologists had suspected ever since the fly-by by the Voyager probes, a quarter-century before: an ice pack, which in places is astonishingly like that one finds in the Arctic ocean,

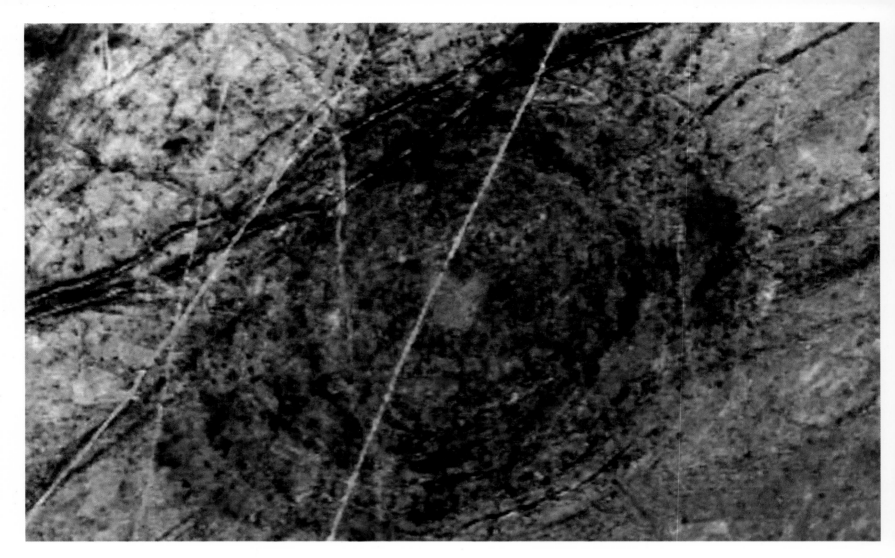

Ganymede and Callisto. The first orbits at a distance of 1 070 000 km from Jupiter, in a perfectly circular orbit, which it completes in slightly more than a week. This satellite of Jupiter definitely merits being considered as a planet in its own right. Its diameter is nearly 5280 km, which places it as eighth in size among the bodies in the Solar System, ahead of Mercury, the Moon, Titan, or even Pluto. Its mass, 148 billion billion tonnes, is half that of Mercury, slightly more than that of Titan, and twice as much as that of our Moon. Ganymede's density, 1.94, is that of a fairly light body, consisting of a mixture of various ices and rock. Orbiting more than one million kilometres from Jupiter, Ganymede is probably little affected either by the giant planet, or by the two innermost satellites, Io and Europa. In fact, its surface, which is not protected by any atmosphere, reveals immense regions covered in impact craters that are as old as the oldest cratered plains on the Moon or Mars. Some of the craters are surrounded by brilliant radial patches, a sign that the bodies that smashed into the surface exposed the pure, subterranean ice. But Ganymede also conceals geological treasures, which have the specialists puzzled. These curious grooved terrains cover millions of square kilometres with more-or-less parallel furrows and ridges, which intertwine and intersect in all directions, with no apparent logic. These vast networks of grooves drawn in Ganymede's icy plains, several hundreds of kilometres long, may reach heights of 1000 m. They are peppered with craters, so their origin, which is still unknown, is extremely ancient.

Callisto, the last of Jupiter's giant moons, is also the least surprising of these small worlds. Callisto takes nearly seventeen days to complete its circular orbit, which lies about 1 880 000 km from Jupiter. Its diameter, 4800 km, and its mass, 108 billion billion tonnes, mean that it is slightly smaller than Gany-

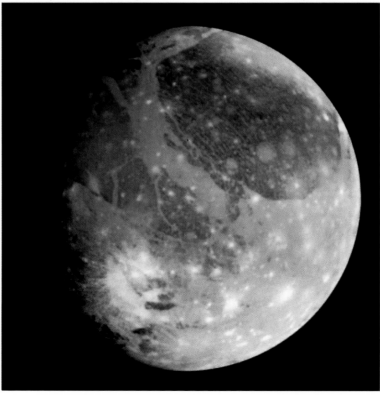

■ GANYMEDE, HERE PHOTOGRAPHED BY THE GALILEO SPACEPROBE, IS THE LARGEST AND THE MOST MASSIVE OF ALL THE SATELLITES IN THE SOLAR SYSTEM. ITS VERY COMPLEX SURFACE REVEALS A LONG, RICH, GEOLOGICAL HISTORY. PART OF THE BODY IS COVERED IN CRATERS, WHILE CERTAIN REGIONS HAVE NONE, BUT INSTEAD EXHIBIT A VAST NETWORK OF LINES AND PARALLEL OR INTERLACED BANDS, WHOSE ORIGIN STILL REMAINS UNKNOWN.

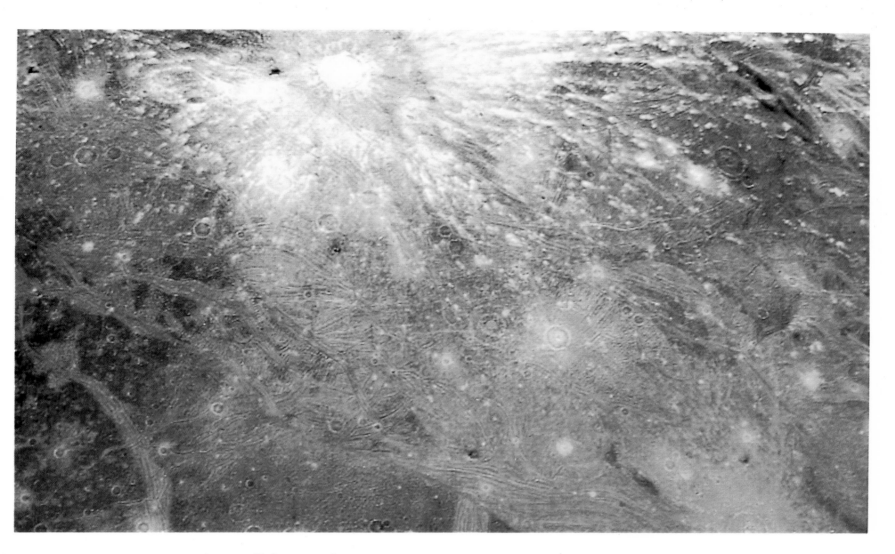

■ CLOSE-UP OF GANYMEDE. THIS AREA, NAMED GALILEO REGIO, IS A PLAIN DOMINATED BY A LARGE AND VERY BRIGHT CRATER. THE IMPACT, UNDOUBTEDLY RECENT, HAS EXCAVATED THE BRIGHT, SUBTERRANEAN ICE, AND SCATTERED IT ACROSS THE SURFACE. RIGHT ACROSS GALILEO REGIO THERE ARE NETWORKS OF PARALLEL LINES.

By measuring the way in which Europa's gravitational perturbations affected the Galileo probe during its close approaches, planetologists have determined the depth of the ocean on Europa as several tens of kilometres. That implies a volume of water that is twice or three times as great as all the oceans on Earth! With regard to the ice on the surface, Galileo was not able to measure its thickness accurately. From the youth of the ice pack, which is practically devoid of impact craters, and the extent of its movements, planetologists now estimate that, in places, its thickness is no more than between one and a few kilometres. One can fully understand why the story of this satellite of Jupiter inflames the imagination of dreamers and poets, but also of scientists. At the very moment, four billion years ago, that waves were lapping in Europa's giant ocean, life appeared in the seas on Earth. Could such a chemical and biological process have also occurred on Europa? And if

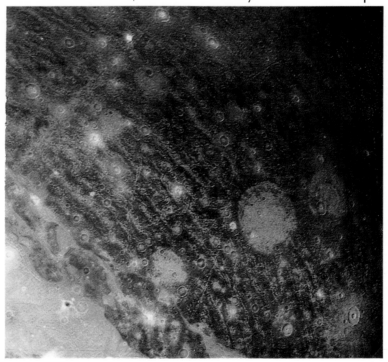

■ THE OLDEST TERRAINS ON GANYMEDE REVEAL THIS SMALL BODY'S CATACLYSMIC PAST. SEVERAL GENERATIONS OF IMPACT CRATERS ARE VISIBLE HERE. THE SMALLEST AND BRIGHTEST, WHICH CLEARLY BREAK THROUGH THE ICY CRUST, ARE THE MOST RECENT. BUT MUCH OLDER CRATERS ALSO APPEAR HERE, AS WELL AS LARGE, ALMOST CIRCULAR, DARK SPOTS.

so, could this extremely hypothetical life still survive today? It is difficult not to think that, beneath their coating of ice, the deep abysses on Europa resemble the cold, dark waters of the Arctic Ocean, hidden beneath its pack ice during the long terrestrial winter. To pierce the ice on Europa, and the mysteries of its deep ocean, American planetologists have conceived the Europa Orbiter mission, financing of which is currently being studied by NASA. If the project does go ahead, the Europa Orbiter probe could leave Earth in 2006 and go into orbit around Europa between 2007 and 2009. Once there, its powerful cameras could map the surface of this most intriguing of Jupiter's satellites in detail. Its radar, above all, capable of penetrating the ice, would finally reveal the secrets hidden beneath the greatest ice pack in the Solar System.

Beyond the mysterious plains of Europa, there are two more large satellites of Jupiter to be visited:

■ CALLISTO IS THE LAST OF THE FOUR LARGE GALILEAN SATELLITES. LITTLE INFLUENCED BY JUPITER'S GRAVITATIONAL PERTURBATIONS, THIS BODY HAS HARDLY EVOLVED FOR OVER THREE BILLION YEARS, AND ITS CRUST IS COVERED WITH IMPACT CRATERS. ITS SURFACE, LIKE THAT OF GANYMEDE, BASICALLY CONSISTS OF ICES AND SILICATE ROCKS.

mede. Its density, 1.86, is again slightly less than that of the largest satellite in the Solar System, and like it, Callisto has an internal structure and a surface strongly influenced by its composition of silicates and ices. This is a dead world, whose landscape has probably not changed for more than 3.5 billion years. Its surface is saturated in impact craters. Here there are no mountain chains, no volcanoes, no traces of lava flows, nor even ancient glaciers emerging from beneath the surface. Just billions of impact craters. Just like the four Galilean satellites, of which the spaceprobes have so far given us only an incomplete picture, albeit in greater and lesser detail, our knowledge as a whole of the miniature planetary system formed by Jupiter and its satellites remains very patchy, and fails to answer many questions. The maps that the American planetary scientists have painstakingly prepared of Io, Europa, Ganymede, and Callisto are marred in places by large irregular areas, as empty and white as the paper on which they are drawn. These are the regions that the spaceprobes have yet to photograph. The Solar System still contains *terrae incognitae* to explore.

not far from the North Pole, when break-up occurs. But this is a vast, limitless ice pack, some 30 million square kilometres in extent: the greatest in the whole Solar System. Under Galileo's sharp eyes, the fine, sinuous lines discovered by the Voyager probes have been revealed as being of incredible complexity. It is, in fact, a breathtaking tracery of faults that intersect one another at all angles, and which, in places, run right across vast shattered areas consisting of large plates thousands of square kilometres in extent. The latter have obviously moved relative to one another, and have even, occasionally, rotated. Certain fractures, which run straight as a die for hundreds of kilometres suddenly change direction when they cross these chaotic zones. We can see that we are dealing with true icebergs, which have slowly drifted around. It is even possible to estimate their height, thanks to their shadows cast on the surrounding surface. The average is 100 to 1000 metres.

So this confirms the scientists' intuitive idea that the surface of Europa is geologically active. The major surprise provided by the Galileo probe, however, was that the crust of ice that covers Jupiter's satellite is probably much thinner than had previously been supposed. Because the ice pack has been broken up in this way, it has undoubtedly been regularly subjected, either in the past or even today, to stress by the swell of the ocean that is hidden beneath the shell of ice.

Saturn:

the Lord of the Rings

■ SATURN'S SYSTEM IS ONE OF THE MARVELS OF THE SOLAR SYSTEM. ITS RINGS OFFER A SIGHT OF UNSPEAKABLE BEAUTY, BUT THEIR DELICATE, UNSTABLE EQUILIBRIUM REMAINS A SCIENTIFIC ENIGMA. HERE, THE PLANET IS VISIBLE THROUGH THE CASSINI DIVISION, AND THE RINGS ARE CASTING THEIR GIGANTIC SHADOW ONTO THE GIANT GASEOUS WORLD. ABOVE THE PLANET, THE SATELLITE TETHYS APPEARS AGAINST THE BLACK BACKGROUND OF THE SKY.

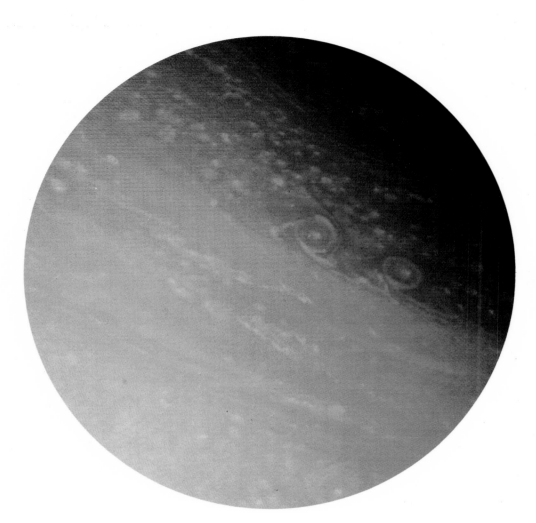

■ LIKE JUPITER, SATURN IS A WORLD THAT IS BASICALLY GASEOUS, MAINLY
HYDROGEN AND HELIUM. THIS VERY-HIGH-RESOLUTION IMAGE SHOWS THE
CLOUDS THAT CIRCULATE AROUND THE PLANET'S NORTH POLE.

arkness. We are drifting through dense clouds of hydrogen and helium, carried by a rising current of irresistible force. The air here is thirty times as dense as the Earth's atmosphere, and the temperature is about 20 °C. Lifted up as if by some giant hand, we reach a region that is less dense, consisting of a yellowish haze, penetrated by the pale light of the Sun. The strength of the wind passes all understanding: an anemometer would show gusts of 1800 km/h. Nowhere else in the whole Solar System are there such violent winds. Suddenly, the clouds part; in a perfectly black sky a brilliant star appears, illuminating a dozen crescent moons. The break in the clouds grows with frightening speed, revealing a monstrous, sparkling arc, hanging suspended above the clouds, which race past, faster and faster. An utterly amazing revelation. One might liken it to a gigantic rainbow, straight from the Land of Oz, that has somehow materialized like magic in our own world. It crosses the entire sky, from east to west. It consists of an infinite number of concentric bands, here sparkling like fresh ice, there as tenuous as a veil of crepe, and farther on so transparent that glimmers of starlight may be seen through it. This breathtaking spectacle of indescribable beauty has no equal on any other planet. We are in the upper atmosphere of Saturn, at an altitude where the pressure (1 bar) is the same as that on Earth. The temperature, by contrast, so far from the Sun, is frigid: just −135 °C.

The clearing in the upper atmosphere expands at an extraordinary speed, and soon extends to thousands, and then millions of square kilometres.

Suddenly seized by a powerful descending current, we are plunged back into night and lose sight of Saturn's rings. We shall not observe how the break in the clouds that allowed us to see the amazing spectacle of the rings enlarges even more. But the sight did not escape eyes on Earth. Henceforward they would know that, every thirty years, the high cloud veil in the upper atmosphere of Saturn clears, revealing the violent hurricanes that affect the lower layers. These major atmospheric changes on Saturn are linked to the planet's closest approach to the Sun. The last occurrences of these gigantic summer storms on Saturn were in 1903, 1933, 1960, and 1990. The next is therefore awaited around 2020.

After Jupiter, Saturn is the largest planet in the Solar System both by mass and by size. It orbits at an average distance of

■ THE FOUR GAS GIANTS IN THE
SOLAR SYSTEM ARE SURROUNDED
BY RINGS. THOSE OF SATURN ARE,
HOWEVER, BY FAR THE BRIGHTEST
AND MOST IMPRESSIVE. HERE,
LARGE DARK PATCHES SEEM TO
DARKEN THE RINGS. THESE ARE
METEORITIC DUST RAISED
ABOVE THE PLANE OF THE DISK
BY THE PLANET'S POWERFUL
MAGNETIC FIELD.

1.43 billion kilometres from the Sun. At such a distance, its orbital velocity is low, 9.64 km/s, and the ringed giant takes some thirty years to complete an orbit. Because the latter is slightly eccentric, Saturn's distance from the Sun in fact varies by some 150 million kilometres between perihelion and aphelion. With its mass of 569 000 billion billion tonnes, the planet has 28 per cent of Jupiter's mass, but almost 100 times that of the Earth. Above all, however, it has the lowest density of any planet in the Solar System. Its density – that is, its mass per unit volume – is just 0.70. In comparison, the average density of the Sun is 1.4, of Jupiter 1.34, and of Neptune and Uranus, the other two gas giants in our system, 1.73 and 1.30, respectively. Given that the reference density, that of water, is 1.0, the conclusion is self-evident: if there were an ocean large enough to take it, Saturn would float!

Saturn may be low in density, but it is still gigantic. Its diameter of 120 000 km is ten times that of the Earth. The globe, which is basically gaseous, rotates extremely fast, in just over ten hours, and is very flattened. The reduction at the poles is close to 10 per cent. Seen from Earth through a telescope, Saturn is like a jewel in the sky. With its small yellowish globe surrounded by its famous brilliant rings, it is a fascinating sight. The image of this somehow unreal world has inspired many to become astronomers and astronauts. Seen in this way from a great distance, from astronomical observatories on Earth, the planet seems, like Jupiter, to be covered in a series of cloud bands, parallel to the equator. In contrast, however, these bands have a far lower contrast and their structure is less complex than those seen on Jupiter.

■ These two photographs were taken by the Hubble Space Telescope in August (top) and in November 1995, during the short period when Saturn's rings were seen edge-on. In the top image, Titan is causing an eclipse on the planet. Below, Dione and Tethys appear, one on each side of Saturn's globe.

Like Jupiter, Saturn formed from the elements that were most abundant in the primordial nebula, the vast disk of gas and dust that gave birth to the Sun and its retinue of planets. Between them, Jupiter and Saturn represent nearly 99 per cent of the mass of all the planets. Like Jupiter, Saturn consists primarily of hydrogen and helium, but its atmosphere also contains, in trace quantities, methane, ethane, ammonia, acetylene, ethylene and phosphine.

When one plunges into Saturn's atmosphere, it becomes much denser and one is soon surrounded by total darkness, split from time to time by lightning. The temperature and pressure increase smoothly: at a depth of about 100 km, the temperature has already passed 0 °C, and lower down, the physical conditions are too extreme for any instruments built by humans. In any case, no space mission to visit the clouds of Saturn is currently planned. At a depth of a few thousand kilometres, the atmosphere transforms into a sort of superdense, searingly hot, tar-like mixture. At a depth of 30 000 km, this gaseous 'ocean', stronger than steel, and at a temperature of 8000 °C, is subjected to a pressure that exceeds 1 million bars. Even deeper, we would find – if we could actually undertake such a trip – a temperature close to 15 000 °C, where the pressure (which is rather an abstract concept at such values) is more than 10 million bars. Here, we would be at the planet's core. Twice as large as the Earth (because scientists say that it has a diameter of about 30 000 km), this core – rocky like those of the other planets – must have an enormous mass, equivalent to between ten and twenty times the mass of the Earth.

Like Jupiter's system, which mimics the solar retinue, Saturn's system is rich and complex. Currently, at least thirty satellites are included, some as large as planets, such as Titan, others as small as asteroids, such as Atlas. But, above all, Saturn is bedecked with a staggeringly beautiful necklace: a magnificent system of rings, about whose amazing complexity we discover more and more every year. They have intrigued astronomers ever since their discovery, nearly four centuries ago. They are divided into three zones that differ in brightness, streaked by a dozen darker divisions, and lie precisely in Saturn's equatorial plane. Because the latter is inclined at 27° to the planet's orbital plane, the rings appear at different inclinations when viewed from Earth. They seem to open and close over a 15-year cycle, i.e., half an orbit. When they are at maximum opening, they frame the planet and, when they are viewed precisely in their plane, they appear like a very fine, almost invisible thread – which is proof of their extreme thinness. It has been known for a long time that they do not revolve as a single body, but as a myriad particles, separated by considerable distances from one another, and each behaving like an independent mini-satellite.

In fact, the rings of Saturn would pose an almost insoluble problem to any astronaut/explorer who might want to find the best angle to observe and photograph them. Seen from a great distance, at several million kilometres, from a spacecraft – such as Voyager 1 and 2, which encountered them in 1980 and 1981, or Cassini, which will approach them early in the 21st century – and fully illuminated by the Sun, they appear magnificent, but show few details. Observing them from the planet is unfortunately impossible, because probably no one will ever send manned probes to surf Saturn's dreadful storms. Choosing a neighbouring satellite, such as Janus or Mimas, as the site for an observatory might, at first sight, seem an excellent idea, but it isn't. Lying, like them, in the equatorial plane, from the satellites the rings appear no more than a fine, sparkling line crossing Saturn and the sky! The problem is the same within the rings themselves: they are so thin that it is practically impossible to see them in perspective. If a spacecraft were suddenly to appear inside them, its occupants would find themselves in a sort of cloud of hailstones, moving in all directions.

■ ON SATURN, THE ATMOSPHERIC
CIRCULATION TAKES PLACE PARALLEL
TO THE EQUATOR, WITH THE
VIOLENCE OF THE WIND
DECREASING AS ONE APPROACHES
THE POLES. THE SPOTS VISIBLE IN
THIS PHOTOGRAPH MOVE AT JUST
60 KM/H ALONG THE TROPICAL
CLOUD BELT. AT THE PLANET'S
EQUATOR, BY CONTRAST, THE
VELOCITY OF THE WIND MAY
REACH 1800 KM/H.

■ This magnificent photograph of Saturn was taken by the Hubble Space Telescope before the planet was disturbed by the great storm of 1990. The angle of inclination of the rings relative to Earth is almost at its maximum. The details visible on this photograph are detectable when the planet is observed with the world's largest telescopes.

Saturn's rings are immense: the rings known as Ring A, Ring B, and Ring C, which are the brightest and the only ones readily visible from Earth, extend from 76 000 to 137 000 km from the centre of the globe. Ring D, which is very faint, extends nearly down to the cloud-tops. Finally, beyond the extremely tenuous Rings F and G, Ring E extends out to a distance of almost 500 000 km from Saturn. This last ring is therefore larger than the orbit of the Moon. As well as being immense, Saturn's rings are also extraordinarily thin: their thickness is between 100 and 1000 m! If you can imagine a sheet of paper, 400 or 500 metres across, you will get an idea of how thin the rings are, compared with their extent.

The appearance of the rings changes completely, depending on whether they are seen in reflected light at different angles and directions, or against the light. Ring B, for example, the densest, reflects sunlight intensely, because its reflecting power – its albedo – approaches 70 per cent. By contrast, it is completely opaque when it is backlit. Just the opposite applies to Ring C, appropriately called the Crêpe Ring because of its gossamer appearance, which consists of smaller particles that are more thinly distributed. It appears very dark in reflected light, but bright with scattered back light. Saturn's rings undoubtedly consist in part of a fine ice and rock dust, and in part of blocks of rock some tenths of a

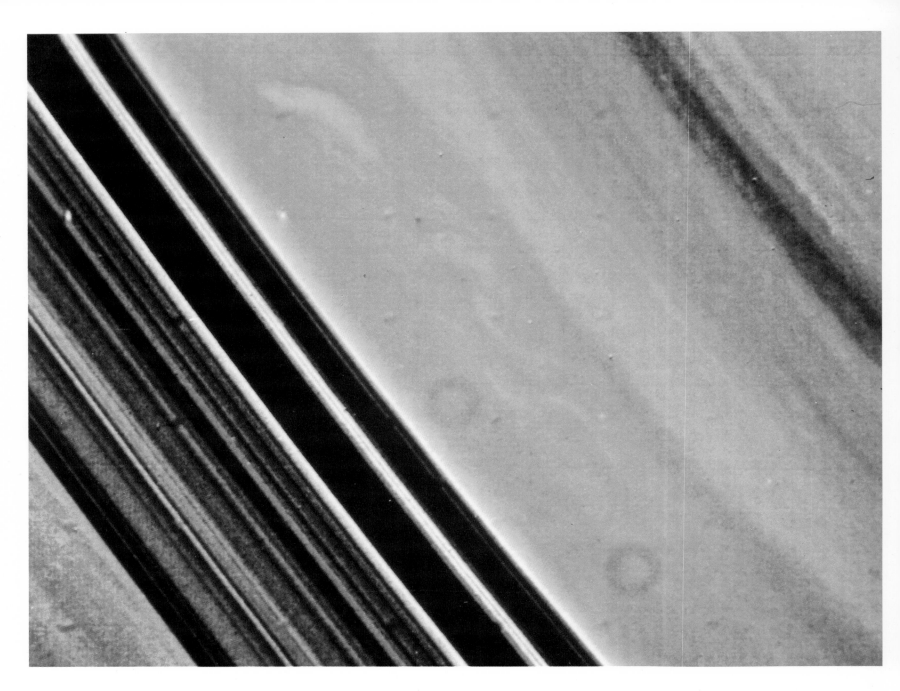

■ SATURN'S RINGS LIE PRECISELY IN THE PLANET'S EQUATORIAL PLANE. THEIR THINNESS IS EXTRAORDINARY, BECAUSE THEIR THICKNESS IS ESTIMATED AS AMOUNTING TO JUST A FEW HUNDRED METRES! ON THIS IMAGE, THE RINGS, SEEN AT A LOW ANGLE, APPEAR VERY DARK AGAINST THE BRILLIANT GLOBE OF THE GIANT PLANET.

metre in diameter, covered in frost.

The ring system, which, when seen from a distance appears so simple, smooth, and regular, proves to be of extraordinary complexity when viewed from close by. For example, the three rings (A, B, and C) visible from Earth, are resolved, in the Voyager images, into several thousand concentric ringlets. The gravitational influence of Saturn's satellites has been responsible for altering the density and composition of the rings, such that in certain zones there is a high particle density, and in others very much less. Large satellites, such as Mimas and Enceladus, which orbit just outside the system, have completely denuded certain orbits. Their perturbations create extremely precise stationary waves throughout the disk, where material accumulates. So the rings, far from being diffuse, appear exceptionally sharp, as if cut out by a razor. Other, much less prominent satellites, such as Pandora and Prometheus, are known as 'shepherd satellites' because of their effect within the system. These two small bodies have orbits that are very close to one another, just 2000 km apart! They literally circumscribe the F Ring, which has a radius of 140 210 km, with a width of just a few tens of kilometres. All round their orbits, Pandora and Prometheus brush past this ring and perturb it, which, like a fragile ribbon, oscillates, deforms, and is distorted. In places, the gravitational tides become so complex that the ring divides into five or six ringlets, which rejoin farther on, or appear to intertwine like snakes!

The interactions between Saturn's satellites, the planet itself, and its rings do not stop there. The last of Saturn's rings (the E Ring), the widest, and also the least dense of all – and which extends, we recall, out to a distance of nearly 500 000 km from the planet – encompasses the orbits of seven satellites and is markedly denser and brighter in the neighbourhood of the satellite Enceladus. According to the experts, the E Ring is 'fed' with material from the icy satellite. Enceladus, either through volcanic eruptions (which have, however, not yet been observed) or through meteoritic impacts, may eject particles of ice into space, and thus feed the ring. Finally, the body of Saturn itself, which has a very intense magnetic field, also exerts a large-scale influence on the system. The magnetic storms that affect the planet also perturb the smallest particles, which may be displaced by the magnetic field's lines of force. Forced slightly above or below the plane of the rings, these levitated particles form long, narrow

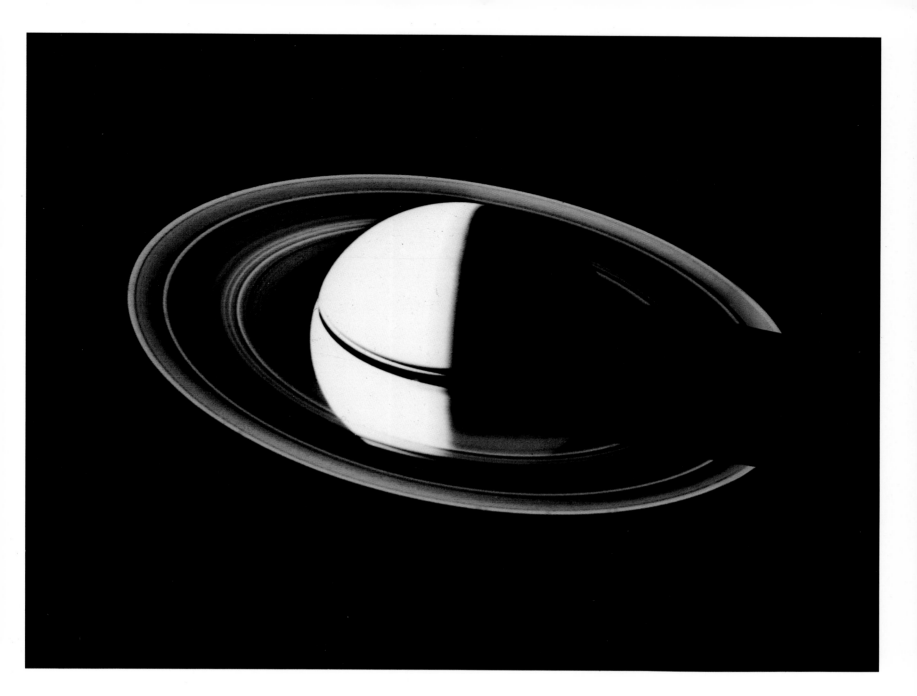

■ THE HIDDEN FACE OF SATURN'S RINGS IS PERHAPS THE MOST SPECTACULAR. AS IT RECEDED FROM THE GIANT PLANET, VOYAGER 2 CAPTURED THIS BACKLIT IMAGE. HERE, THE RINGS NO LONGER REFLECT SUNLIGHT, BUT DIFFUSE IT INSTEAD. IN FACT, THE DENSEST RINGS APPEAR DARKEST, WHILE THE CASSINI DIVISION BECOME EXTREMELY BRIGHT.

and ephemeral dark 'spokes' on the face of the disk.

Rings are a feature common to all the giant planets. Although those of Saturn may have been known since 1610, those belonging to Uranus, Jupiter, and Neptune were not discovered until 1977, 1979, and 1984, respectively. Saturn's system of rings is by far the most massive, the largest, the densest, and the brightest. Researchers estimate that the total mass of Saturn's rings is about 10 million billion tonnes, which represents the mass of a satellite like Mimas, which is 400 km in diameter! In comparison, the rings of Jupiter, Uranus, and Neptune must be between one thousandth and one millionth of that mass.

At present, no one knows the origin of planetary rings. The zone in which rings are found around the four giant planets is undoubtedly an important key to resolving the enigma. All the rings occur round about the Roche Limit. Within this 'forbidden zone' surrounding every planet in the Solar System, no satellite can form through accretion, and if an already formed satellite ventures in, it is subject to enormous gravitational tides, which eventually cause it to fragment. In either case, we would therefore witness the formation of a system of rings around the planet. So, are the rings of Saturn, and its giant siblings, vestiges of the formation of the planets, and is their age therefore 4.5 billion years? If so, how could a system so strange and complex as Saturn's disk of rings survive for so long? If not, then were the rings formed in the much more recent past, by chance encounters between asteroids and comets with the giant planets?

Titan:

an Earth in hibernation

■ THE MOST ENIGMATIC WORLD OF THE WHOLE SOLAR SYSTEM ORBITS SATURN. TITAN IS PERPETUALLY HIDDEN BENEATH A THICK ATMOSPHERE, AND WE HAVE NO KNOWLEDGE WHATSOEVER OF THE CONDITIONS THAT PREVAIL AT ITS SURFACE. THE ATMOSPHERE OF TITAN, LIKE THAT OF THE EARTH, PRIMARILY CONSISTS OF NITROGEN. BUT, SO FAR FROM THE SUN, THE SMALL WORLD IS PLUNGED INTO EXTREME COLD. WHAT CHEMICAL REACTIONS CAN TAKE PLACE AT A TEMPERATURE OF $-180\,^{\circ}$C?

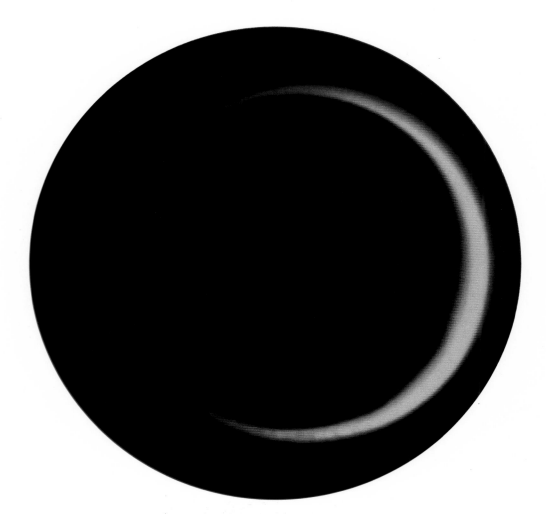

■ TITAN'S UPPER ATMOSPHERE, PHOTOGRAPHED AGAINST THE LIGHT BY VOYAGER
2. LIGHT FROM THE SUN, SCATTERED IN ALL DIRECTIONS BY THE HAZE, FORMS A
MOCK CRESCENT, WHICH NEARLY ENCIRCLES THE WORLD.

We are far away from Saturn. Far away from the Sun as well, and from its warm inner planets, whether welcoming or desert. Seen from here, at a distance of more than one million kilometres, Saturn resembles a giant, yellow child's balloon, around which a forgotten hoop is spinning… The Sun, whose tiny disk is difficult to detect, at a distance of nearly one billion five hundred million kilometres, shines with just one per cent of the brightness that it has on Earth. But even from the distant orbit of Titan, the steady, blinding brilliance of our star forces any observer to look away. The world over which we are flying is mysterious: It is really a small planet – although it cannot truly be given that name, because, strictly speaking, it is a satellite of Saturn – that is surrounded by a thick atmosphere. Its surface, like that of Venus, is completely hidden beneath a uniform, monotonous blanket of clouds.

As we orbit the body, we pass into shadow. Seen from here, Titan, plunged into utter darkness, should show nothing but a disk as black as ink, outlined against a sky full of stars. But not so; the sight that meets our eyes has a beauty and a strangeness that take our breath away. The planet's atmosphere, which scatters sunlight, creates a luminous crescent that almost completely encircles the night-time disk. This body is truly puzzling: far from the Sun and the terrestrial planets with familiar surfaces, it travels alone, like some small lost Earth, right in the middle of the gas-giant planets. Where did Titan come from?

After Ganymede, which orbits Jupiter, Titan is the largest of the satellites in the Solar System. In size, it is ninth in the Sun's retinue; by mass, the tenth. Its diameter of 5150 km far exceeds that of the Moon (3476 km), of Pluto (2400 km), and is also just greater than that of Mercury (4880 km). Although larger than Mercury, Titan is, however, much lighter. The mass of the planet, which formed from heavy metals, not far from the Sun, exceeds 330 billion billion tonnes, whereas that of Titan is less than 140 billion billion tonnes. Like the giant or icy bodies that occupy the outer Solar System, Titan was thus formed primarily from ices and light gases, and its density is less than 1.90. Its orbit is nearly circular and lies in the same plane as Saturn's equator and rings. It takes sixteen days to complete one orbit, at a distance of slightly more than 1.2 million kilometres from Saturn, i.e., three times the Earth–Moon distance.

■ TITAN'S UPPER ATMOSPHERE, AS
IT APPEARS BETWEEN ALTITUDES OF
200 AND 300 KM. THE
ATMOSPHERE BECOMES OPAQUE AT
AROUND AN ALTITUDE OF 230 KM.
THE ORANGE CLOUDS CONSIST OF
NITROGEN, METHANE, AND MORE
COMPLEX MOLECULES. HIGHER,
A LAYER OF BLUISH HAZE,
WHICH IS FAR LESS DENSE,
SCATTERS SUNLIGHT, LIKE THE
SKY ON EARTH. AT THIS
ALTITUDE, THE
TEMPERATURE IS
−100°C.

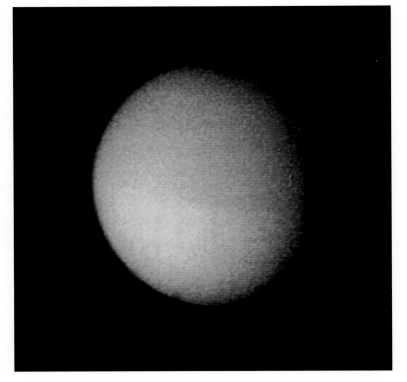

■ UNFORTUNATELY, TITAN'S ATMOSPHERE IS OPAQUE AT OPTICAL WAVELENGTHS, AND SPACEPROBES HAVE NEVER BEEN ABLE TO PHOTOGRAPH THE BODY'S SURFACE. WE WILL NOT FINALLY KNOW THE TRUE FACE OF THIS TINY, HIBERNATING PLANET, UNLESS THE DIFFICULT CASSINI-HUYGENS MISSION SUCCEEDS. THIS AIMS TO LAND A PROBE ON THE SURFACE, AND THIS IS EXPECTED TO OCCUR IN 2004.

Titan is a unique world, perhaps the most fascinating in the whole Solar System. In Saturn's system, it is the sole giant satellite, because the other twenty-nine known satellites of the gas giant are between one hundredth and one hundred millionth of its mass. Titan has monopolized more than 90 per cent of the available mass, including the rings, that exists around Saturn! Its size and its orbital parameters do not allow there to be any doubt about its origin: it was formed at the same time as the Sun and the planets, some 4.5 billion years ago, and is one of the original bodies in the Solar System. Titan is also the smallest body in the Solar System to have retained a very thick atmosphere, like the Earth or Venus. Compared with that of Titan, the atmospheres of Pluto, Triton, or Io are no more than almost non-existent wisps of gas. Why is Saturn's large satellite swathed in fog and clouds in this way, while Mercury, the Moon, Ganymede, Europa, and Callisto, for example, are just barren deserts exposed to the rigours of space? We don't know. One indication, however, at least enables us to explain how Titan has been able to retain a significant gaseous envelope. Its distance from the Sun is twice that of the satellites of Jupiter, ten times more than that of the Moon, and twenty-five times that of Mercury. At Saturn's distance, at which it condensed, the violence of the solar wind was therefore weaker, and the temperature of the interplanetary medium was much colder. Gases and ices, instead of subliming and disappearing, as on most of the closer planets and satellites, remained on Titan. On Saturn's largest satellite, the molecules of gas are less excited by solar radiation, and consequently never reach the crucial velocity of 2.5 km/s that would allow them to escape into space. In its planet's calm, frigid environment, at more than 1 billion kilometres from the Sun, Titan was able to retain its atmosphere for more than 4.5 billion years, whereas Mercury, the Moon, and Ganymede were not.

The atmosphere of Titan, like that of the Earth, primarily consists of nitrogen. This inert gas forms 78 per cent of the terrestrial atmosphere, and more than 90 per cent of Titan's. That does not mean, of course, that an Earthling could safely go and explore the surface of Titan: our essential gas, oxygen, is completely absent! By way of contrast, however, the satellite's clouds contain methane as well as nitrogen, and possibly argon and several rarer chemical compounds, including numerous hydrocarbons.

Can anyone imagine what sort of landscape Titan's clouds are hiding? From Earth, all that is visible through a telescope is a tiny, blurred disk, and the two Voyager probes that skimmed past it in 1980 and 1981 were unable to pierce the layer of bluish- and orange-coloured hazes that form a uniform cover. On the other hand there is no lack of radar soundings, spectra, gravimetric and photometric measurements to be used in theoretical models, which are all constrained by the known physical conditions: mass, density, chemical composition, gas pressure, etc. Researchers now know that Titan's upper atmosphere lies at altitudes of between 200 and 300 km. At this high altitude, where the Sun still shines, the temperature nearly reaches −100 °C. The pressure in this uppermost layer, which primarily consists of methane and ethane, is less than 1 millibar. Lower down, as one approaches the surface – and unlike what happens in the atmospheres of Jupiter and Saturn, which are heated by their internal energy sources – the temperature decreases, while the pressure increases and solar illumination decreases. At the actual surface of Titan the pressure is extraordinarily high for such a small body: 1.5 bar as against 1 bar at the surface of the Earth. The temperature, in the absence of any sunlight, is terribly low: about −180 °C. On Earth, during the cold of the polar night everything freezes: the ice pack, rivers, steppes, and tundras. On Titan, where oxygen does not exist in its free form, and where water, as hard as steel, is locked into the subsoil in the form of ice, methane and nitrogen possibly play the role of active liquids. At the pressure and temperature found there, these two gases, which are so difficult to maintain in liquid form on Earth, are actually close to their triple point, at which they are able to exist simultaneously in solid, liquid, and gaseous states. But there are seasons on Titan, because of Saturn's orbit, which is 1.35 billion kilometres of the Sun at its closest point, and 1.5 billion kilometres at its farthest. During the four seasons, each of which lasts around eight of our years, the temperature on Titan may vary, depending on latitude, between −170 °C and −210 °C. As we shall see later, that could create some

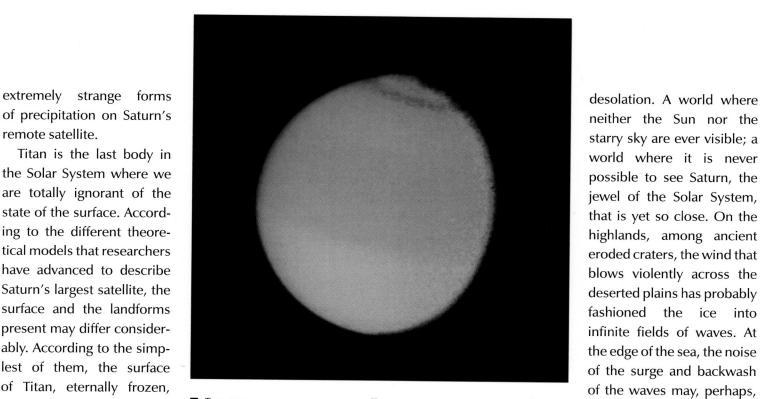

■ THE CLOUD LAYER THAT SURROUNDS TITAN IS NOT PERFECTLY UNIFORM. ON THIS PHOTOGRAPH, A SLIGHT DIFFERENCE IN TONE IS DETECTABLE BETWEEN THE NORTHERN AND SOUTHERN HEMISPHERE. IN ADDITION, THIS IMAGE SHOWS A VERY DARK, BUT MARKED, CLOUD FEATURE THAT ENCIRCLES THE BODY'S NORTH POLE.

extremely strange forms of precipitation on Saturn's remote satellite.

Titan is the last body in the Solar System where we are totally ignorant of the state of the surface. According to the different theoretical models that researchers have advanced to describe Saturn's largest satellite, the surface and the landforms present may differ considerably. According to the simplest of them, the surface of Titan, eternally frozen, resembles that of Triton, Neptune's satellite. There is an icy crust, with occasional craters and mountains, covering the rocky core of the small world. Some researchers, however, suggest that this icy crust does not cover the rocky core itself, but instead an enormous ocean of methane. This second hypothesis therefore likens Titan to Europa. The surface of this giant satellite would be a vast ice sheet covered with clouds. In October 1994, the Hubble Space Telescope finally managed to see through Titan's cloud layer at infrared wavelengths. Bright and dark patches, whose nature remains unknown, were discovered. Only one thing is certain: the surface of Titan is not uniform.

Finally, the last model proposed to account for Titan's surface sensibly incorporates all the earlier theories. Titan might well have continents of varying glacial relief, whose shores are washed by the waves formed on vast seas of methane. It is to finally discover what conditions actually occur on this hibernating world that American and European scientists conceived the Cassini–Huygens mission, which left Earth on 15 October 1997. The objective of the Cassini probe, designed by NASA, is to examine Saturn's system from top to bottom. As for the Huygens module, built by the European Space Agency (ESA), it is due to descend into Titan's atmosphere and land on the surface. It should provide us with the most fascinating planetary exploration that we have ever undertaken.

While we await the arrival of Huygens on Titan, in 2004, we can still dream of the possible science-fiction landscapes that it may reveal. A world permanently plunged into polar night, veiled in fogs, illuminated as if by some invisible moon, and blasted by frigid snow showers. Shadows and utter desolation. A world where neither the Sun nor the starry sky are ever visible; a world where it is never possible to see Saturn, the jewel of the Solar System, that is yet so close. On the highlands, among ancient eroded craters, the wind that blows violently across the deserted plains has probably fashioned the ice into infinite fields of waves. At the edge of the sea, the noise of the surge and backwash of the waves may, perhaps, cause our astronaut to forget for an instant that this liquid, with its silver reflections, is not water … In winter, when the temperature drops below −200 °C, an incessant rain of nitrogen falls on the grey desert, and the sea slowly becomes covered in methane icebergs, which form ominous silhouettes through the veil of fog that obscures the horizon.

Strangely, this world that we imagine to be totally frozen by the cold is actually extremely active. In Titan's upper atmosphere, where the Sun still penetrates through the clouds, nitrogen and methane are dissociated by its ultraviolet radiation, and then recombine into molecules of greater and greater complexity, which subsequently slowly fall towards the surface. An incredibly rich chemistry is taking place, at an extraordinarily low temperature, synthesizing heavier and heavier molecules: ethane, butane, propane, ethylene, and acetylene. Even at the surface, these hydrocarbons occur as brown, oily puddles scattered over the icy plains. These pools contain even more complex molecules, whose names are full of meaning for biologists: hydrogen cyanide, cyanogen, cyanoethylene. These molecules, DNA bases, are the building blocks of life.

Some 4 billion years ago, the Earth, like Titan, was swathed in a nitrogen atmosphere, devoid of oxygen, and rich in methane. It was these same methane products that, in the warm terrestrial oceans, enabled the appearance of organic molecules, then of amino acids, and finally of the first living cells. As far as Titan is concerned, it has been plunged into the most frigid of winters for 4.5 billion years. But what has happened on this genuine, small, hibernating Earth during its long polar night?

Enceladus

and the worlds of ice

■ WITH THE EXCEPTION OF TITAN, SATURN'S SATELLITES
ARE SMALL-SIZED BODIES WITH LOW DENSITIES.
ENCELADUS, WHICH IS PROBABLY THE MOST INTERESTING
AND THE MOST MYSTERIOUS OF THESE WORLDS,
ESSENTIALLY CONSISTS OF ICE. ITS DIAMETER IS JUST
500 KM. DEVOID OF ANY ATMOSPHERE, IT EXHIBITS
TERRAINS THAT HAVE FEW CRATERS, OR ARE EVEN
PERFECTLY SMOOTH, A SIGN OF SIGNIFICANT
GEOLOGICAL ACTIVITY.

■ A CLOSE-UP OF THE SURFACE OF RHEA. THE SURFACE OF THIS MODERATE-SIZED
SATELLITE OF SATURN IS SATURATED WITH IMPACT CRATERS, LIKE CERTAIN AREAS OF
THE MOON. RHEA IS LESS REFLECTIVE THAN ENCELADUS. THE ICE ON ITS SURFACE
IS MIXED WITH A DARKER MATERIAL.

The view doesn't change, but it is impossible to tire of it. On this airless world, the sky is eternally black, and it is not always easy to distinguish night from day. Here, the starry heavens are always visible and the Sun, some 1.5 billion kilometres away, is no more than a large star, with no apparent disk, but whose brilliance is intolerable. Any astronauts who visit this strange body one day will not be able to see the world from which they came. Too close to the Sun, too far away from here, the Earth is invisible. Beneath their feet, the explorers will discover ice, frozen in the utter cold, as strong as steel. Nights on Enceladus are terrible: before dawn, the temperature can drop to nearly − 220 °C. On this satellite of Saturn, however, the days are also frigid, because, in full sunlight, at midday, the thermometer never exceeds − 170 °C.

Even equipped with heavy survival equipment and bottles of oxygen, an explorer would hardly weigh more than about 2 kg, because of this world's feeble gravity. A jump would cause them to rise a dozen metres in the air, where they would be able to see beyond the horizon, which is less than 1 km away on this tiny body. A few gentle kangaroo bounds, and our third-millennium astronaut would cover several hundred metres, flying, with elegant parabolic leaps, over a landscape of strange beauty. Everywhere there would be the same arctic scene: plains of dazzling ice, with gentle slopes, or covered in absolute labyrinths of giant crystals with prismatic faces; glacial valleys with steep, mirror-smooth slopes. All around, sunlight is reflected back by the brilliant ice, making the landscape glitter with fleeting stars. Farther away, rise mountains of ice, which sparkle as the Sun moves across the sky. On this vast ice sheet, punctuated here and there with craters excavated into the bright ice, our astronaut will be able to search for the numerous meteorites littering the ground, which are easy to spot, being as black as coal, and broken into a thousand pieces by the violence of the collision with the surface of Enceladus.

Enceladus is one of the wildest and most beautiful worlds in the Solar System. It also offers an exceptional vantage point for seeing Saturn's system. Like all the satellites, Enceladus is gravitationally 'locked' by the giant planet, which it orbits in thirty-three hours. Because this orbital period is equal to the time it takes to rotate on its axis, the satellite continually points the same side to the planet. As a result, from one whole hemisphere of Enceladus, Saturn appears absolutely immobile, hanging motionless above the landscape, while the starry sky slowly turns behind it. Seen from the polar regions of the small planet, emerging above the mountains visible on the southeastern horizon, Saturn presents its most majestic appearance. Probably no one has ever imagined such a scene: the ringed planet appears as an enormous crescent, rising high in the sky. Part of the globe, plunged into darkness and visible only because it blocks the view of the stars, crackles with violent, silent storms, transient sparks, quickly absorbed by the fantastic cloudy mass of Saturn. Its rings, seen precisely edge-on, form a blinding line, as straight as a die, and unimaginably thin.

Enceladus is extraordinarily close to Saturn. At a distance of 238 000 km, the sight of the ringed planet is enchanting, immense, brilliant, and almost unnatural. At night, Saturn replaces the Sun and radiates enough light to illuminate the landscape. Everywhere, the ice glimmers in the soft yellow light, which casts shadows that are strangely diffuse and blurred. The apparent diameter of Saturn is more than 25°, which is fifty times that of the Moon as seen from Earth!

To the northwest, there is a sight that is nearly as beautiful and puzzling. In a black sky, five coloured crescents can be seen, perfectly aligned with one another. They do not remain fixed in the sky of Enceladus, unlike Saturn, but, in a slow, stately ballet, each follows its own unique path. Brilliant Tethys, twice the size of the Moon as seen from Earth, will have disappeared below the

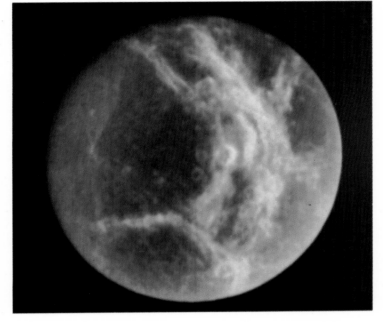

■ DIONE, LIKE MIMAS, ENCELADUS, TETHYS, OR RHEA, ALWAYS TURNS THE SAME SIDE TOWARDS SATURN. THIS PHOTOGRAPH SHOWS DIONE'S HIDDEN SIDE, I.E., THE SIDE ON WHICH ONE WOULD NEVER SEE SATURN. VERY BRIGHT ALBEDO MARKINGS STREAK THE SATELLITE. THEY PROBABLY REPRESENT FRACTURES THROUGH WHICH BRIGHT ICE HAS ESCAPED ONTO THE SURFACE.

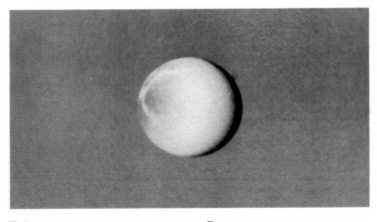

■ THIS SPECTACULAR IMAGE OF THE SATELLITE DIONE PASSING IN FRONT OF THE DISK OF SATURN WAS RECORDED BY THE VOYAGER 1 SPACEPROBE. THIS SCIENCE-FICTION IMAGE WAS REPEATED WITH MOST OF SATURN'S SATELLITES, INCLUDING TITAN, BUT IS ALMOST IMPOSSIBLE TO OBSERVE FROM EARTH.

■ SATURN'S SATELLITES ARE TINY WHEN COMPARED WITH THE GIANT PLANET. HERE, THE SATELLITE DIONE, WHICH MEASURES 1120 KM IN DIAMETER, PASSES SLOWLY IN FRONT OF THE GIGANTIC GLOBE OF GAS. SATURN IS NEARLY 1 MILLION TIMES AS MASSIVE AS DIONE.

horizon in a few hours, together with smaller Dione and Rhea, which will mutually eclipse one another before sinking in their turn. Titan, 1 million kilometres away, merely appears as a brownish-coloured crescent, half the size of our own Moon. Tomorrow, at the same time, our astronaut will find it in practically the same place in the sky of Enceladus, but its earlier temporary companions will have disappeared, lost somewhere on their orbits on the other side of Saturn.

Saturn currently has thirty known satellites, with orbital parameters and physical characteristics that are more or less well-known. Its sphere of influence, which seems to be the richest and most complex of all the Solar System, undoubtedly extends to numerous other small bodies, mini-satellites lost in the middle of its rings, and which we shall recognize only after the exhaustive, minute examination that the Cassini–Huygens mission is due to carry out in a few years' time. As we have seen in the last chapter, Titan is the largest and most massive of Saturn's satellites, and none of the other bodies in the system reach even 2 per cent of its mass. The smallest satellites of Saturn (Pan, Atlas, Prometheus, Pandora, Epimetheus, and Janus) all orbit very close to one another, at distances of between 133 580 km and 151 472 km from the centre of the planet – in other words, at the edge of the brilliant A Ring. They are no more than large irregular rocks, measuring between 20 and 100 km in diameter.

Saturn's 'classical' system of satellites – known since the 18th century – consists of small, true planetary bodies, spherical in shape and with varied, mutable surfaces. These are Mimas, Enceladus, Tethys, Dione, Rhea, Titan, and Iapetus, in increasing order of distance from Saturn. These bodies have the lowest densities found in the Solar System and – surprise, surprise – mainly consist of ice, surrounding a tiny rocky core. The two smallest bodies, and the closest to Saturn, Mimas and Enceladus, have diameters of 390 km and 500 km, respectively. They have record low densities of 1.17 and 1.24, which may be compared with that for water, which is 1.00. The mass of Enceladus, 84 million billion tonnes, is thus one thousandth of that of the Moon. Saturn, for its part, is more than 5 million times more massive than its tiny satellite. Tethys and Dione, the next satellites, are larger; 1050 km and 1120 km, but scarcely much denser: 1.26 and 1.44, respectively. Their masses are also greater, because Dione, for example, amounts to almost exactly 1 billion billion tonnes. Finally, Rhea and

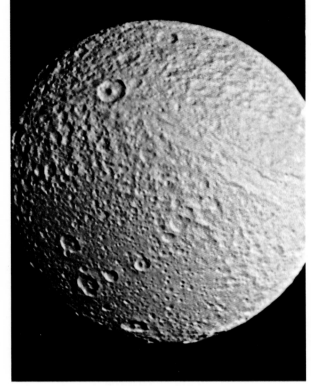

■ TETHYS EXHIBITS ONE OF THE MOST ANCIENT SURFACES IN THE SOLAR SYSTEM. ALTHOUGH SATURATED WITH CRATERS, IT DOES, HOWEVER, SHOW TRACES OF GEOLOGICAL ACTIVITY. THE SATELLITE IS, FOR EXAMPLE, GASHED BY A GIGANTIC CANYON, ITHACA CHASMA, WHICH MEASURES MORE THAN 2000 KM LONG.

Iapetus, the most distant, are even larger. Iapetus has a diameter of 1440 km and a density of 1.21. Rhea, which is the largest satellite of Saturn after Titan, measures 1530 km across, with a density of only 1.33 and a mass of 2.5 billion billion tonnes – less than 2 per cent of the mass of Titan!

Finally, a third family of satellites orbits Saturn at very great distances. Hyperion and Phoebe are just large irregular blocks, some one hundred kilometres across. Phoebe takes one-and-a-half years to complete one orbit around Saturn, at a distance of nearly 13 million kilometres. From the point of view of celestial mechanics, Saturn's system exhibits (in addition to the exceptional richness of its association of rings and satellites) some true oddities, which are not found anywhere else in the Solar System. Around Saturn, several satellites occupy the same

■ MIMAS IS PERHAPS THE SMALLEST BODY IN THE SOLAR SYSTEM THAT MAY BE CONSIDERED A PLANET IN ITS OWN RIGHT. DESPITE A DIAMETER OF JUST 390 KM, THIS WORLD DOES, IN FACT, EXHIBIT A TRULY SPHERICAL SHAPE, WHICH IS GENERALLY THE PREROGATIVE OF MUCH LARGER BODIES. MIMAS IS COVERED IN IMPACT CRATERS.

orbit! Atlas, Prometheus, and Pandora occupy orbits that are less than 4000 km apart. Epimetheus and Janus are effectively in the same orbit. Luckily the laws of gravitation prevent them from ever encountering one another. Tethys, for its part, is accompanied by two mini-satellites, Telesto and Calypso, which follow and precede it, respectively, in its orbit. Finally, Dione also has a guardian angel, Helene, a small, irregular body some 50 kilometres across. Have these small satellites, trapped on apparently dangerous and unstable orbits, been captured recently; are they debris from violent collisions between the larger bodies; or have they been orbiting this way ever since the Solar System was formed? We simply do not know.

In the extreme cold that prevails at a distance of 1.5 billion kilometres from the Sun, the surface of Saturn's satellites, despite being formed of ice, reacts to meteoritic impacts like rock. In fact, when seen from a distance, these small worlds circling Saturn resemble Mercury, the Moon, or Callisto. Like those bodies, most of Saturn's minor satellites appear covered in craters. Mimas, Tethys, Dione, and Rhea are practically saturated with them. The surfaces of these distant worlds therefore bear witness to the fact that the evolution of our planetary system has been the same throughout, whatever the distance from the Sun; whether a few million or a few billion kilometres. We see their surfaces today just as they appeared 4 billion years ago.

Enceladus, however, has a topography that does not tally with the general model. This small, icy satellite, a tiny body relative to most of the bodies in the Solar System, should, in principle, be geologically inactive. Only the larger planets, such as the Earth or Mars, for example, have a sufficiently large mass of rock to lead to internal heating – caused either by the radioactivity inherent in some of

■ THE SATELLITE DIONE IS HEAVILY CRATERED. ITS SURFACE IS THEREFORE EXTREMELY ANCIENT, AND THIS IS ALSO SHOWN BY THE VIOLENCE OF SOME OF THE IMPACTS. THE LARGEST CRATERS VISIBLE IN THIS IMAGE EXCEED 100 KM. LIKE THE LARGE LUNAR CRATERS, THEY SHOW A DISTINCT CENTRAL PEAK, DESPITE THE FACT THAT HERE THE SURFACE OF THE BODY LARGELY CONSISTS OF ICE.

their constituent elements, or by the enormous pressure that reigns in their interiors, or by both at once. Plate tectonics and folded mountain chains are evidence of our own planet's inner activity, just as the volcanoes and canyons indicate the same for Venus and Mars. Another source of potential heat is to be found in the powerful gravitational tides that act on Io, for example. This tiny body is overflowing with energy, and releases it in utterly unbridled volcanic activity.

Enceladus, which is tiny and very light, does not satisfy any of these conditions that are needed to be geologically active, yet it is one of the most active bodies in the Solar System. Although it is partially cratered, its surface has no impact craters with diameters larger than about 25 km, proving that the earlier collisions, which were the most significant ones, have been obliterated. By comparison, Mimas is scarred by a crater 130 km in diameter. The surface of Tethys has traces of an enormous crater 400 km across, and Dione and Rhea has several craters of between 100 and 200 km. But what was a major surprise to astronomers was to discover that a large part of the surface of Enceladus is smooth. There is not the slightest crater for tens of kilometres in all directions. These icy areas seem to be extremely young. The ground reflects light like fresh snow. The average albedo for Enceladus is more than 90 per cent, the highest in the Solar System. In places, ancient, highly cratered plains have been flooded by the material that has risen to the surface. Researchers estimate the age of these glacial valleys as less than 100 million years. Why and how has ice spread out on the surface of Enceladus? Perhaps, like Europa, Enceladus has an ocean beneath its icy crust. The glaciers that spread out over the surface may therefore arise from geysers that have erupted warm water under pressure from

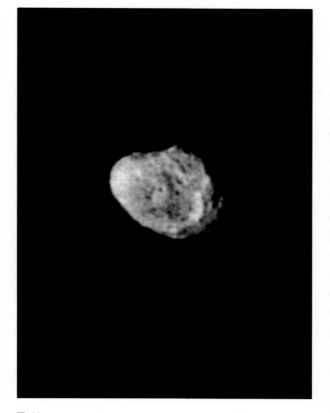

■ HYPERION ORBITS AT A GREAT DISTANCE FROM SATURN, AND ITS CHARACTERISTICS ARE STILL POORLY KNOWN. ALTHOUGH ITS MASS IS ESSENTIALLY THE SAME AS THAT OF MIMAS, HYPERION DOES NOT HAVE THE SPHERICAL SHAPE THAT IS CHARACTERISTIC OF PLANETARY BODIES. ITS DIMENSIONS ARE APPROXIMATELY $400 \times 260 \times 200$ KM.

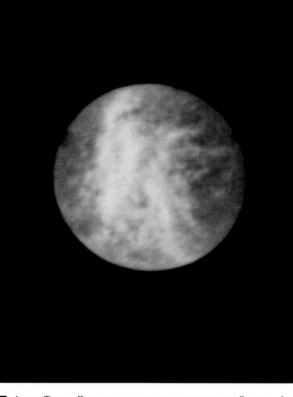

■ AFTER TITAN, RHEA IS THE LARGEST SATELLITE OF SATURN. ITS SURFACE IS ALMOST EVENLY COVERED WITH CRATERS. ON THIS IMAGE, RHEA APPEARS FULLY ILLUMINATED BY THE SUN. THE CRATERS ARE INVISIBLE, BUT IN CONTRAST, THE BRIGHT, WHITE ICE IS CLEARLY VISIBLE.

beneath the surface, and which would have frozen immediately it touched the ground. Unless the crust of Enceladus, like an ice sheet, is deformed and cracked from time to time, under pressure from its hidden ocean. We therefore do not know whether some particular form of volcanism exists on Enceladus; a volcanism that turns out to be strange and exotic, where the volcanoes are perfect cones of ice, and where the lava is water, initially liquid, and then solid. Certain specialists suspect, however, that Enceladus feeds the E Ring, in which it orbits, with ice crystals. It is true that on such a small, light world, it would be easy for a geyser to put water into orbit. The escape velocity for Enceladus is about 700 km/h, one fifteenth of that for the Earth.

Saturn's system, with its rings, its mini-satellites that undergo complex interactions, Titan (a fascinating world in hibernation), and its strange satellites of ice, still conceals numerous mysteries. Explaining these will help us to better understand the origin and evolution of the Solar System as a whole. This is the ambitious objective of the Cassini–Huygens mission. Be patient, the American–European spaceprobe will not reach the neighbourhood of the ringed giant until 2004.

Uranus:

a recumbent giant

■ URANUS IS THE MOST POORLY KNOWN PLANET IN THE
SOLAR SYSTEM. TOO DISTANT, IT SHOWS NO VISIBLE
DETAILS IN TERRESTRIAL TELESCOPES. VISITED ONLY BY THE
VOYAGER 2 SPACEPROBE, IN 1986, URANUS HAS
REMAINED HIDDEN BENEATH A THICK LAYER OF BLUISH
HAZE, AND WE KNOW NOTHING OF ITS ATMOSPHERIC
PHENOMENA. LIKE JUPITER, SATURN, AND
NEPTUNE, THIS GAS GIANT PRIMARILY CONSISTS
OF HYDROGEN AND HELIUM.

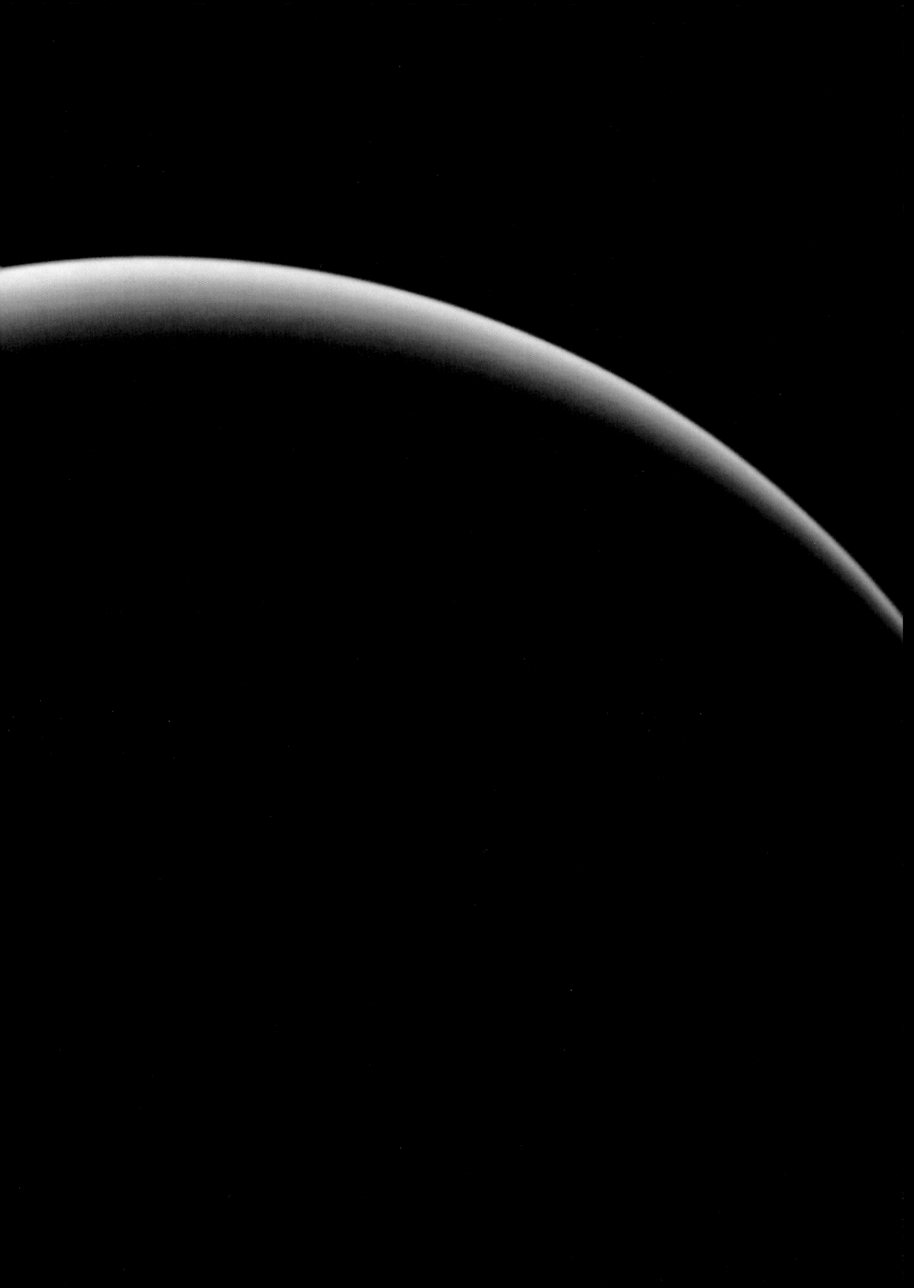

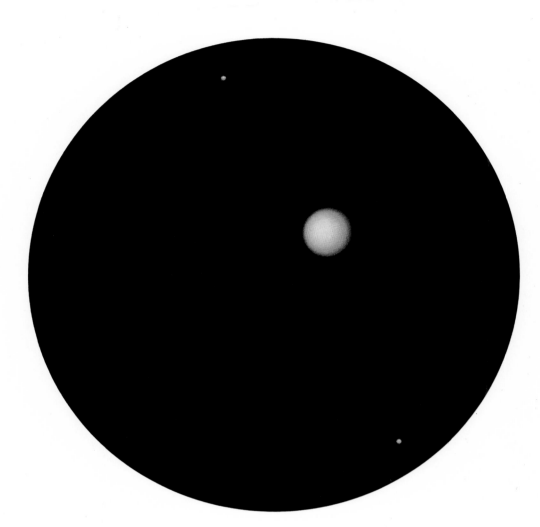

■ Uranus is at the limit of naked-eye visibility. When observed from
Earth with a powerful telescope, the giant planet reveals only a small
bluish disk, accompanied by a few tiny points of light, which are its
largest satellites.

Seen from here, the blue planet appears enormous, standing out against the completely black sky as a brilliant disk more than 110° in diameter. The highest wisps of cloud on the giant planet seem almost within touching distance, even though 50 000 km away. We are on a tiny body, Cordelia, which orbits Uranus about 25 000 km from the planet. This tiny, irregular satellite some fifteen kilometres across, completes an orbit in just eight hours. Perpetually fixed in Cordelia's sky, Uranus never changes. Here there are no turbulent, highly coloured clouds as on distant Jupiter, but a globe that is uniformly diffuse and bluish in colour.

Unlike the Sun, the Moon, Mercury, Venus, Mars, Jupiter, and Saturn, bodies known ever since the first human beings lifted their eyes towards the sky, Uranus does not belong to the ancient history of humanity, nor to its mythology. As seen from the Earth, in fact, this distant body, essentially point-like, is at the limit of visibility to the naked eye. It was only, in fact, by chance that the amateur astronomer William Herschel discovered Uranus on 13 March 1781. Visible only as a tiny bluish disk in even the world's most powerful telescopes, this body revealed practically nothing about its physical charac-

teristics to astronomers. Until 1986, when the Voyager 2 spaceprobe encountered the giant planet, we knew essentially nothing about it, apart, of course, from its orbital characteristics, which had been carefully measured for more than two centuries.

Uranus orbits at an average distance of 2.8 billion kilometres from the Sun, on a nearly circular orbit, which it completes in slightly more than eighty-four years. It is the least massive of the four gas giants in the Solar System. With a mass of 87 000 billion billion tonnes, Uranus is only slightly less massive than Neptune, but is about fifteen times more massive than the Earth. The giant planet's diameter is slightly more than 50 000 km, and its density is almost 1.30. Uranus is a world that is basically gaseous, primarily consisting of hydrogen, helium, methane, and ammonia. Its upper atmosphere, like that of Titan, is covered by an almost completely uniform layer of haze, which prevents us from seeing the major climatic changes, the storms, and the motions of the clouds on the globe. Scientists assume that there is a hot, rocky core at the centre of Uranus, which is subject to a pressure of several million bars. The uniqueness of Uranus, however, lies in its

orbital motion. For most of the planets the equatorial plane is precisely, or at least approximately, parallel to the mean plane of the Solar System – the ecliptic. The equator of Mercury, like that of Jupiter, is within 3° of the ecliptic. The equator of the Moon is inclined at less than 7°, while the inclinations of the equators of the Earth, Mars, Saturn, and Neptune lie between 23 and 30°.

Uranus, however, has a peculiarity all its own: its equatorial plane is, in fact, more-or-less perpendicular to its orbit! This produces some extraordinary seasonal effects. Every forty-two years, one of the two poles of the planet points almost directly towards the Sun, which is then nearly at the zenith. For half of a Uranian year, which corresponds to forty-two terrestrial years, one polar region is perpetually plunged into darkness, while the other polar region has continual daylight. The layer of haze that covers the planet prevents us from seeing what specific atmospheric conditions arise as a consequence of this extraordinary orientation of the planet, which is perhaps the result of some ancient catastrophe. The impact of a giant planetoid at a particular angle may perhaps explain the change in the planet's orientation.

Like the three other gas giants, Uranus has an impressive satellite system. No less than twenty satellites are now known to orbit the giant planet. A large number of them, such as Cordelia, Ophelia, Bianca, Cressida, and Desdemona, are no larger than the satellites of Mars. Juliet, Portia, Belinda, and Rosalind, with diameters of between 30 and 60 km, are slightly larger. Puck measures 75 km. All these small satellites were discovered by Voyager 2 in 1986, whereas the five 'historical' satellites of Uranus, which are much larger, were discovered between 1787 and 1948. These are, in order of increasing distance from the planet, Miranda, Ariel, Umbriel, Titania, and Oberon. In size and density, these five objects are comparable to the small satellites of Saturn. Miranda measures

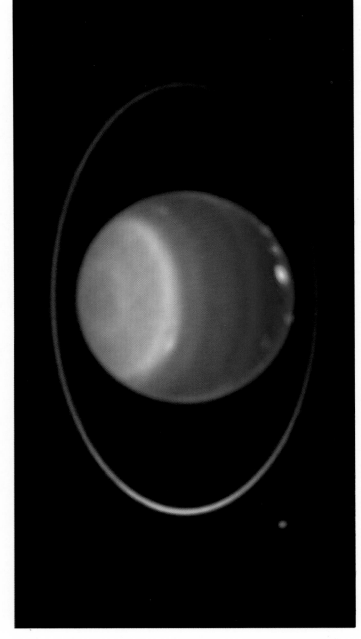

■ THIS FALSE-COLOUR PHOTOGRAPH OF URANUS AND ITS SYSTEM OF RINGS WAS OBTAINED BY THE HUBBLE SPACE TELESCOPE IN AUGUST 1998. A FEW HIGH-ALTITUDE CLOUDS ARE SEEN IN THE PLANET'S ATMOSPHERE. THE RINGS, VERY NARROW AND OF IRREGULAR BRIGHTNESS, AS WELL AS SOME OF THE TWENTY SATELLITES OF URANUS (WHOSE BRIGHTNESS HAS BEEN ARTIFICIALLY INCREASED), ARE ALSO CLEARLY VISIBLE IN THIS IMAGE.

only 470 km in diameter; Ariel and Umbriel, about 1200 km; Oberon, 1500 km; and finally, Titania, 1600 km. This last body, with a density of 1.7 and a mass of 3.5 billion billion tonnes, has roughly the same characteristics as Rhea.

Miranda is the best known of all the satellites of Uranus, thanks to the very high resolution images obtained by the Voyager 2 spaceprobe. The object, despite being small, was revealed to be extremely interesting. The topography of this world is one of the most fantastic in the whole Solar System. In one region we have thousands of square kilometres of a fairly dark terrain, consisting of a mixture of rock and ice, and which is heavily cratered, as if we were on the Moon or Mercury. Another region has arrays of absolutely parallel ridges and furrows that form perfect geodesic lines following the strongly curved surface of the tiny world. In these areas, the impact craters have practically disappeared. The geological features found on Miranda defeat the specialists. In the region known as the 'Chevron', two series of furrows meet but run together at more-or-less a right angle, with no apparent reason. Adjoining this enigmatic region, an enormous cliff rears a vertical wall, 15 000 m high, above jumbled terrains, devastated by some cosmic catastrophe.

How could such an incredible geology arise on such a small body? The thermal energy stored within the core cannot possibly account for Miranda's geography. The effects of various gravitational couples acting between the body, Uranus, and its other satellites, do not serve any better. The powerful gravitational tides that have literally melted the interior of Io are no more than gentle ripples at Miranda's location. On second thoughts, however, the satellite does give the impression of having suffered an extremely violent shock in the distant past. A theoretical scenario may even be advanced. Miranda, nearly 4.5 billion years ago, was partially dismantled as the result of the impact of a large planetoid. Whole fragments of the satellite were detached, and may well have floated alongside the body for some time before being drawn back by its gravitational field. At that epoch, Miranda, only just having formed, was probably still hot and malleable, which would have made it easier for the missing pieces to be 'stuck back together'.

Less spectacular than Miranda, Ariel is nevertheless rich in information about the Uranian system. At first sight, its surface seems to be saturated with impact craters, apparent proof of a great age and of an evolution

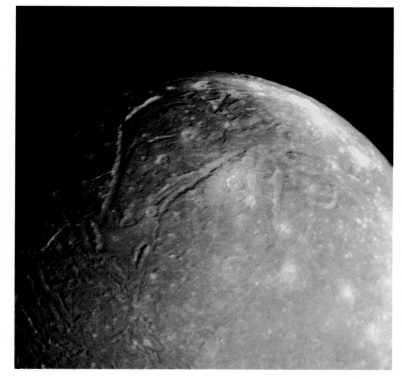

■ WITH A DIAMETER OF 1180 KM, ARIEL IS ONE OF THE LARGER SATELLITES OF URANUS. LARGELY COVERED WITH IMPACT CRATERS, THIS SMALL BODY ALSO EXHIBITS MUCH RARER AND MORE INTERESTING GEOLOGICAL FEATURES: LONG, DEEP VALLEYS THAT SCORE CERTAIN AREAS FOR HUNDREDS OF KILOMETRES. THEY MAY HAVE RESULTED FROM ANCIENT EXTRUSIONS OF ICE ONTO THE SURFACE.

that ceased early in the Solar System's history. Yet the only craters seen on Ariel are small ones. The first generation of craters, excavated by the large planetoids that roamed the early Solar System, has been completely obliterated. And finally, the satellite is scored by a remarkable system of valleys, several hundreds of kilometres long, the base of which apparently shows no signs of cratering. Ariel's surface has

therefore been extensively reworked. But we still have no idea of the process involved. Some planetologists suppose that when the Uranian system originated, the orbits of the satellites may have been chaotic and, generated gravitational waves that were sufficiently intense to heat some of their number. Today, the five major satellites of Uranus calmly orbit the giant planet. Their orbits are almost perfectly circular and in precisely the same plane as the equator! Miranda takes thirty-four hours to orbit Uranus at just 100 000 km above the highest clouds. Oberon, in contrast, completes one orbit in thirteen-and-a-half days, at a distance of more than

550 000 km. Apart from being endowed, like Jupiter, Saturn, and Neptune, with a major satellite system, Uranus also has a system of rings. These are far less dense than those of Saturn, and are very dark. They were not discovered until 1977. Like Saturn's rings, they are immense – the Epsilon Ring has a diameter of 100 000 km – and extraordinarily thin, because their thickness is estimated to be less than 100 m.

In the first analysis, the satellite system of Uranus resembles a miniature version of the systems of Jupiter and Saturn. The planet itself is about one tenth of Saturn's mass and, from the point of view of mass and size, lies between the terrestrial planets, such as Venus and the Earth, and the gas giants, like Jupiter. In the same way, the satellites of Uranus are about one tenth of the mass of Titan or the Galilean satellites. Their density, which is significantly greater than Saturn's icy satellites, does not cease to amaze the experts, who expected

to find that as one gets farther from the Sun, the bodies would have lower and lower densities. Throughout the universe, the same physical laws produce the same astronomical structures everywhere: clusters, galaxies, nebulae, stars, and, finally, planets. When examined in detail, however, we find that small differences in local conditions and, above all, the chance factors that affect the fate of each individual body have resulted, at least in the case of our own planetary system, in a profusion of different worlds, with complex, and sometimes peculiar, evolutionary histories, as is clearly shown by the example of Miranda.

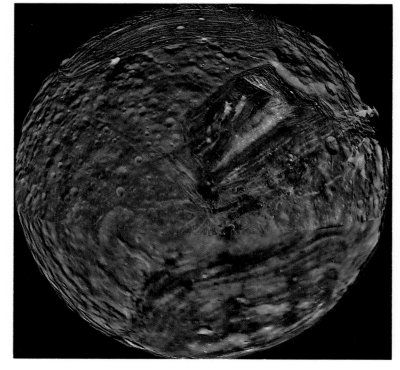

■ MIRANDA IS ONE OF THE MOST ENIGMATIC WORLDS IN THE SOLAR SYSTEM. ITS SURFACE SHOWS THE RESULT OF UPHEAVALS THAT ARE OUT OF SCALE WITH ITS SMALL SIZE. GIGANTIC CLIFFS AND BREATHTAKING CREVASSES BEAR WITNESS TO A CATACLYSMIC PAST. SCIENTISTS BELIEVE THAT MIRANDA WAS ONCE BROKEN INTO SEVERAL PIECES, WHICH WERE THEN 'STUCK BACK TOGETHER' THROUGH THE FORCE OF GRAVITY.

Halley:
the great traveller

■ No other comet has been studied with so many different methods of observation as Comet Halley. During its last return, three spaceprobes were sent to meet it. When Comet Halley is close enough to both the Sun and the Earth it is clearly visible to the naked eye. This image shows the comet in front of the rich star fields of the Milky Way.

■ This is an incredible image. It shows the icy nucleus of Comet Halley for the first time, photographed by the European spaceprobe Giotto. The irregular shape and the violent jets of gas are clearly visible.

The surface of this icy world is porous, cracked, and jumbled, as if it had been subject to intense cultivation. The ground, as black as coal, does in fact largely consist of carbon compounds and organic molecules. Beneath the dark crust, in various places, veins of pure ice sparkle for an instant in the sunlight. But what a pitiful luminary! We are so far from the star that it does not appear as its familiar disk, but simply as a tiny pinprick of intense light. Nonetheless, it is impossible to look at our daytime star for very long; its fixed brilliance soon becomes intolerable and we are forced to look away.

It is unreal: the world on which we stand seems like something out of a dream. The sight is puzzling, and totally incomprehensible. In front of us, a plume of gas and ice has been gushing out in absolute silence for several hours, slowing escaping into space until it mingles with the hazy veils of the Milky Way. All around this geyser, blocks of ice float above the ground, rising and falling in slow motion, in response to invisible waves. Suddenly, the eruption ceases, as fast as it began. A wisp of gas remains for a long moment, like a sheet of morning mist, then sublimes and disappears. Above the crater, which is

now inactive and empty, thousands of shards of ice, clouds of crystals, of flakes, of mineral dust, obstinately remain floating in space, refusing to fall back onto the surface, as they would do on any planet.

We are 3.8 billion kilometres from the Sun, on board the strangest spaceship one could imagine: Comet Halley. This body, so courted by astrologers, divines, and astronomers for more than two millennia, this 'wandering star', this sign of misfortune that has always terrified humanity, is, in fact, a tiny body. Its irregular shape approximates to a rough ellipsoid, whose three axes are $16 \times 8.2 \times 7.5$ km. Comet Halley is therefore the smallest body that we have visited in our trip through the Solar System. Smaller than the asteroid Gaspra ($19 \times 12 \times 11$ km), smaller than Phobos and Deimos, the satellites of Mars, which measure $27 \times 21 \times 19$ km and $15 \times 12 \times 11$ km, respectively. Not only is it small, but Comet Halley is also (and above all) extraordinarily light. It density, which probably does not even reach 0.50, is less than that of Saturn's icy satellites, such as Mimas and Enceladus (1.17 and 1.24), and even less than that of Saturn (0.70). The mass of the comet is, on a planetary scale, infinitesimal: probably less than

TAKEN AT PRACTICALLY THE SAME TIME AS THE IMAGE OF THE NUCLEUS OPPOSITE, THIS PHOTOGRAPH SHOWS THE GIGANTIC COMA AND TAIL THAT COMET HALLEY DEVELOPED. THE DIFFUSE TAIL, EXTENDING FOR MILLIONS OF KILOMETRES, HAS RESULTED FROM THE MATERIAL EJECTED BY THE JETS VISIBLE ON THE CLOSE-UP IMAGE OF THE NUCLEUS. ON THE SCALE OF THIS IMAGE, WHICH WAS OBTAINED FROM EARTH, THE NUCLEUS IS MERELY A TINY POINT.

■ March 1986: Comet Halley crosses the Earth's skies before receding to the edge of the Solar System. Heated by solar radiation, the gas has melted and has been violently ejected from the nucleus. Every second, Comet Halley lost 30

100 billion tonnes! On Earth, that would represent the weight of an ice floe some ten kilometres long. Comet Halley is well and truly a sort of cosmic iceberg, frozen, fragile, and without a rocky core. It is a body that is so light that its gravitational field is negligible. On the comet, astronauts would find that they weigh no more than 1 gram! They would have greatest difficulties in moving about, because the slightest sudden movement would lift them from the surface and would place them in orbit or, even worse, send them out into deep space for ever. The only way of remaining on this body, and of visiting it in utter safety, would be to use an MMU, a Manned Manoeuvring Unit, like those used by astronauts in Earth orbit, when they venture out of the Space Shuttle. The equipment is fitted with small reservoirs full of gas, which when released through directional thrusters, allows the astronauts to move around safely and independently.

Comet Halley, astronomically and strictly speaking, is a planet: like Mercury, Venus, or the Earth, it orbits the Sun – although in the opposite direction to the general motion of bodies in the Solar System. It takes seventy-six years to complete an extremely eccentric, elliptical orbit, which takes it to within 88 million kilometres of the Sun, between Mercury and Venus. Its aphelion – that is, its maximum distance from the Sun – lies beyond the orbit of Neptune, at some 5.3 billion kilometres from our star.

TONNES OF WATER, NITROGEN, AND METHANE. FOR THIS
LONG-EXPOSURE PHOTOGRAPH, TAKEN THROUGH THREE
FILTERS (BLUE, GREEN, AND RED), THE TELESCOPE FOLLOWED
THE COMET'S PROPER MOTION. THIS IS THE REASON FOR
THE UNSHARP, COLOURED IMAGES OF THE STARS.

It was not until compara-
tively late that astronomers
finally understood the true status of the diffuse bodies that
they saw rapidly cross the Earth's sky from time to time. For a
long time, in addition to utterly irrational explanations, which
are of little interest to us here, they seriously wondered
whether they were not dealing with atmospheric phenomena.
The English astronomer whose name was given to Comet
Halley was the first to discover, in the 18th century, that the
bright, hairy objects that had terrorized earthlings for several
centuries were nothing more than one and the same celestial
body. After having observed the spectacular appearance in
1682, he calculated and then predicted its return in 1759. A

triumph for one of the first calculations in celestial mechanics,
the comet returned at the time predicted, although unfortunately sixteen years after Halley's death, who only gained
posthumous glory.

Since then, every seventy-six years, astronomers have
awaited the comet's return. At each of its returns, in fact, it
passes at a greater or lesser distance from our planet. Although,
as we have seen, the object is tiny, it becomes extraordinarily
bright as it approaches the Sun. Halley's surface, which is
exceptionally dark – its albedo is less than 4 per cent – is
extremely efficient in absorbing heat from the Sun. At distances
of less than 450 million kilometres from the Sun, the comet's
ices begin to melt and vaporize beneath the crust. These

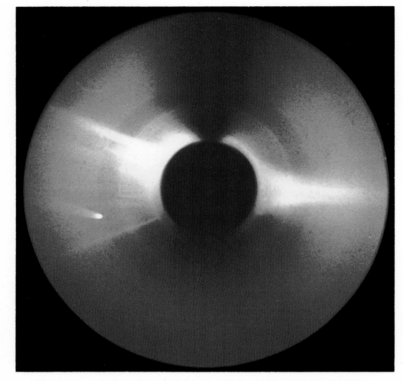

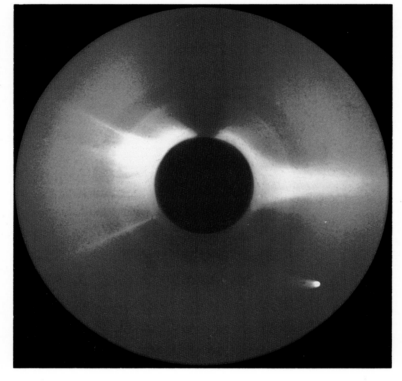

pockets of gas begin to explode, puncturing the surface, and shooting immense geysers out into space, which the feeble gravity is not able to stop. As it plunges towards the Sun, the comet heats up and, under the influence of the particles in the solar wind, the gases emitted by the nucleus are swept outwards, where they first form an elegant head (or coma), which is itself followed by two tails, one consisting of mineral dust and the other of ionized gas, and which may attain the fantastic length of 100 million kilometres! Although essentially a vacuum – a comet's tail does not even amount to one million tonnes – Halley's Comet may well become the brightest object in the night sky.

Close to perihelion, the temperature of the comet's surface exceeds 70 °C. Its activity then becomes paroxysmal, with the gases – mainly water vapour, but also carbon dioxide, nitrogen, methane, and ammonia – being expelled at 500 m/s and at the rate of 30 tonnes per second.

Since its return of 1759, astronomers have followed each of Comet Halley's visits

■ In 1973, Comet Kohoutek passed briefly near the Sun. The astrononauts on board the Skylab space station were able to photograph the event, using a coronagraph, which masks the light from the star. In these two extraordinary images, the comet is shown moving among the coronal rays in the outer atmosphere of the Sun.

with greater consideration and attention. Each new generation of researchers has directed more and more powerful instruments towards this venerable traveller. After its penultimate return, in 1910, it disappeared again from telescopic view in a few months. At the beginning of the 1980s, the hunt for Comet Halley resumed and, on 16 October 1982, the Americans at Palomar Observatory were the first to recover it, as a faint, diffuse trace on an electronic image obtained with the famous 5-m (200-inch) Hale Telescope. The climax of the surveillance of the comet occurred in 1986, when it easily reached naked-eye visibility and was the target of spaceprobes

and thousands of amateur and professional telescopes all over the world. The European Giotto spaceprobe, which passed less than 600 km from the comet, finally obtained a close-up image. After the return in 1986, Comet Halley's brightness slowly faded, as its distance from the Sun increased, its activity diminished, and the number of telescopes able to follow it declined dramatically. In 1991, when the comet, to everyone's amazement, released a sudden puff of gas, it was already beyond the orbit of Saturn, and was visible only in the most powerful telescopes in the world. In 1992, just one of them, at the La Silla Observatory in Chile, was able to photograph it. It should be said that the comet was then around magnitude 26, which means that its brightness was about one-hundred-millionth of the faintest stars visible to the naked eye. But in 1993, Comet Halley was lost. Once and for all? Certainly not. Improvements in the performance of telescopes enabled it to be 'rediscovered' in January 1994. Since then, Comet Halley has been lost once more. But astronomers have not given up hope of being able to recover it before it reaches its maximum distance in the year 2024, more than 5 billion kilometres from the Sun, beyond the orbit of Neptune.

Comet Halley will then turn back and retrace its path, returning to the vicinity of the Earth and the Sun in 2061.

Comet Halley is the most famous member of an enormous family. Humans have already witnessed the passage of several thousands of comets across the Earth's skies, and not only professional astronomers but amateurs as well discover more every year. Because of their orbital and also physical and chemical characteristics, we now know that these minute

bodies condensed, like the planets, 4.5 billion years ago. Their principal component being water, like the satellites of Saturn, for example, we may reasonably assume that they were formed relatively far from the Sun, and from the lightest elements which had been blown towards the outer edge of our planetary system by the primordial star. Like the asteroids, billions of comets crashed into the newly born planets. According to some scientists, repeated collisions of thousands of comets with the Earth may even explain, on their own, the presence of large quantities of water on our planet. Even today, a collision between the Earth and a comet remains a serious menace. A large comet, such as Comet Halley, that crashed onto the Earth would cause a major catastrophe, which might be fatal for half of all living species, possibly including our own.

Once the tumultuous episode surrounding the origin of the planets was over, and some of the comets had been 'swallowed up' by the latter, most of these cosmic icebergs were then ejected out to the fringes of our system by the gravitational slingshot effect of the giant planets. Accurate analysis of the orbital elements of several hundred identified comets enables us to assume that there is a vast reservoir of these bodies at an extremely great distance from the Sun. This toroidal or spherical cloud, first proposed by the Dutch astronomer Jan Oort, should lie about one light-year from Earth – i.e., at about 10 000 billion kilometres. It possibly

■ WHEN COMETS APPROACH
SUFFICIENTLY CLOSE TO OUR
PLANET, THEY MAY BE BRIGHT
ENOUGH TO RIVAL MOONLIGHT,
AND THE TAILS SOMETIMES
STRETCH ALMOST FROM ONE
HORIZON TO THE OTHER. SUCH
SIGHTS ARE RARE. COMET
HALLEY, WHICH HAS BEEN
VISITING EARTH FOR AT LEAST
TWO THOUSAND YEARS, IS
NO LONGER SUFFICIENTLY
RICH IN GAS TO LIGHT UP
THE EARTH'S SKY IN
THIS WAY.

includes 1000 billion comets, which together represent – reckoning 1 to 1000 billion tonnes per comet – a mass equal to that of Jupiter!

The distance of the Oort Cloud is similar to that of the nearest stars, such as Proxima Centauri or Barnard's Star, which are, respectively, 4.3 and 5.8 light-years from the Sun. Because of this, the latter, as seen from a comet in the Oort Cloud, would be indistinguishable from the thousands of other stars that fill the sky. Comets take millions of years to orbit the Sun, and their surfaces, frozen at a temperature of −250 °C, are plunged into eternal night. From time to time, the gravitational perturbations caused by a star passing the Sun's vicinity must disturb hundreds of comets in their orbits, and which then finally escape from the Sun's attraction and wander off through the Galaxy for millions of years, before they are possibly captured by other stars.

Because of these gravitational disturbances, however, other comets do the opposite, and slowly begin to fall in towards the inner Solar System. Over a few hundred thousand years, their velocities increase as the Sun's attraction grows, and then heating of the surface releases the first jets of gas. On Earth, some lucky, and very happy, astronomer snaps it up and has their name assigned to a new comet! Frequently, however, the brilliant illumination of a comet when close to the inner planets signs its death warrant. Although they have lived for 4.5 billion years in the eternal youth of a 'deep freeze', comets become the most fragile and ephemeral bodies when they burn their wings close to the fires of the Sun. Some of them make just a single approach, and then completely disappear, having escaped from the Sun's gravitational field. Others are captured by the Sun, which they thenceforward orbit on normal planetary orbits, albeit strongly elliptical ones. This is the case with Comet Halley, and also with Comet Encke, which came close in 1993, with Comet Schwassmann-Wachmann, which will pass near the Earth in 2004, or yet again with Comet Ikeya–Seki, whose return is impatiently awaited – in 2845! At each perihelion passage these comets develop their long tails of gas and dust, which certainly

■ COMET HALLEY, HERE PHOTOGRAPHED IN APRIL 1986, HAS SUBSEQUENTLY REMAINED VISIBLE IN THE WORLD'S LARGEST TELESOPES. BETWEEN 1990 AND 1994, IT WAS SOMEWHERE BETWEEN THE ORBITS OF SATURN AND URANUS. IT WILL REACH ITS MAXIMUM DISTANCE (APHELION) IN 2024, AND WILL THEN SLOWLY FOLLOW ITS PATH BACK TOWARDS THE SUN.

means that they lose a considerable amount of mass. In the case of Comet Halley, its last return to the vicinity of the Sun caused it to lose more than 100 million tonnes of material or, in other words, a layer of ice, nearly 10 metres thick, from its surface. After several hundred passes low over the surface of the Sun, comets become depleted in their gaseous components and disperse, feeding the vast disk of dust that occupies the Solar System's main plane.

Close encounters between major comets and the Earth are rare – a few per century, at the most – but sometimes leave Earth's inhabitants with an indelible memory. In the case of Comet Halley, it is, of course, the frequency of the encounters between the comet and the Earth that has fired the imagination. More rarely, comets have arrived out of the blue and have remained for several weeks beneath our enchanted gaze. In this respect, the end of the century just past provided us with two magnificent presents. One after the other, at an interval of a year, two splendid comets illuminated the springtime night skies in 1996 and 1997. The first, Comet Hyakutake, literally rushed out of the depths of space, before astronomers had time to prepare to observe it. Discovered on 30 January 1996, its closest approach to the Earth was on the night of 24 March that same year.

For anyone who had the luck to see it, the passage of Comet Hyakutake would remain an unforgettable event. On that clear night of 24 March 1996, the comet's proper motion against the northern constellations was detectable with the naked eye in just a few minutes. It should be said that the body was racing across the Solar System at nearly 150 000 km/h and encountered our planet at a distance of just 15 million kilometres! On that memorable night, the comet, at magnitude 0, was the brightest object in the sky. It had a nucleus that was perfectly point-like, as bright as one of the stars in the Great Bear. The nucleus was surrounded by a coma with a greenish tinge, more than 2° across, or the size of four Full Moons. But it was the tail, which stretched for an incredible distance, that astounded observers. In clear mountain skies, the tail of

Comet Hyakutake could be followed for more than 60°, which is the size of the largest constellations in the sky.

And yet the radar pulses emitted by the powerful 70-m antenna at Goldstone in California, reflected back by Comet Hyakutake, enabled scientists to show that this spectacular comet actually had a minute nucleus, that was just slightly more than 2 km in diameter! Following meticulous measurements obtained during its passage, astronomers have been able to reconstruct the trajectory of the 'Great Comet of 1996'. This confirms that the Comet Hyakutake took 8000 years to complete its last orbit. We shall not see this ancient traveller again as quickly as that, however, because gravitational perturbations to which it has been subjected by the Solar System's planets in recent years have greatly modified its orbit, which will now take it out to a distance of more than 100 billion kilometres from the Sun. The next return of Comet Hyakutake is predicted to be in 17 000 years.

At the end of 1996, while Comet Hyakutake was slowly becoming lost among the stars of the southern sky, another exceptional comet arrived. But whereas Comet Hyakutake was tiny, fast-moving, and rushed across the spring sky more like a meteor, Comet Hale–Bopp made its presence felt a long time in advance, moved across the sky at the slow pace of a spaceprobe and, above all, seemed to be of an utterly exceptional size. This comet was discovered on 23 July 1995, at the record distance of 1 billion kilometres. To be visible at such a distance, two years before it passed close to the Sun and the Earth, the body had to be enormous. Indeed, we now know that Comet Hale–Bopp measures nearly 50 km in diameter with a mass approaching 10 000 billion tonnes. Comet Hale–Bopp, clearly visible to the naked eye, was followed throughout the spring of 1997, both through the polluted and light-polluted skies above cities and through the crystal-clear, rarefied air at mountain observatories. The sight was very different from that shown by Comet Hyakutake. Comet Hale–Bopp's closest approach to the Earth, on the night of 22 March 1997, was only at a distance of 197 million kilometres. In the evening sky, the comet floated, looking unreal, between the

■ Bends in Comet Hyakutake's tail, which consists of gas and dust. In this image, the comet is passing in front of the stars in Ursa Major (the Great Bear); way beyond those, at a distance of 15 million light-years, a beautiful spiral galaxy may be seen.

constellations of Andromeda and Cassiopeia. It appeared much smaller, but also brighter, more compact, whiter, and with a far greater contrast than Hyakutake. Then Comet Hale-Bopp rushed on to its rendezvous with the Sun. On 1 April, it reached perihelion, just 136 million kilometres from the Sun. Subjected to the intense solar radiation, the gas production of the giant comet, which was rotating on its axis in about a dozen hours, reached record proportions. When the frozen surface of the comet was exposed to the heat of the Sun at dawn each morning, powerful geysers, which had been inactive during the night, awoke, releasing 300 tonnes of water per second, and sending enormous plumes of ice into space. For the first time, astronomers were able to see these plumes directly, through a telescope, as they slowly curled round in a spiral, thanks to the rotation of the nucleus. The gas that was released in this way amounted to 25 million tonnes per day and, blown away by the solar wind, formed two vast yellow and bluish tails, which were more than 100 million kilometres long.

In 2000, although the comet was about to leave the neighbourhood of the Solar System, it was still being followed by astronomers. Like Comet Hyakutake, its orbit was greatly disturbed by the enormous mass of the planets that it had encountered in the centre of the Solar System. It will take the comet 1200 years to reach its maximum distance from the Sun, at some 55 billion kilometres. This is an enormous distance, which light, travelling at a velocity of 300 000 km/s, would take two days to cover. Then, slowly, the 'Great Comet of 1997' will curve back, to resume its path towards the nine planets of the Solar System, which it will revisit rather more than one thousand years later.

Nevertheless, many dangers await Comets Halley, Hyakutake, Hale–Bopp and their innumerable icy brethren during their staggeringly long travels. For example, the powerful geysers that escape from their icy surfaces act as reaction motors, capable of producing major changes in their orbits. Comets are often subject to the hazards of perturbed or irregular orbits, and come to a bad end. In 1979, Comet

Howard-Koomen literally fell into the Sun. In 1976, Comet West, which was as fragile as the arctic pack ice at the beginning of spring, was unable to survive its one and only perihelion passage, and broke into several fragments. Finally, at the beginning of 1993, Comet Shoemaker-Levy disintegrated after it was captured by Jupiter. One year later, inexorably drawn towards the giant planet, it crashed into the upper atmosphere. Overall, the life expectancy of a comet in a short-period solar orbit is estimated at just a few hundred million years, which is just the blink of an eye on a cosmic scale.

Ever since the last return of Comet Halley, in 1986, scientists have dreamed of repeating the exploit of the European spaceprobe Giotto, which in March of that year approached to within less than 600 km of the comet, before losing the use of its camera to a sand-blasting by particles of ice and meteoritic dust. It was recovered from hibernation, however, and went on to send back data from the vicinity of Comet Grigg-Skjellerup, which it approached to within 200 km in July 1992. If all goes well, however, the beginning of the 21st century should offer fine opportunities for visiting these great cosmic travellers. Comet Halley is, of course, currently too far away to be visited, so astronomers have set their hearts on short-period comets, which are never very far from the Sun. On 6 February 1999, the small Stardust probe left Earth for a slow trip to Comet Wild 2, which it will reach on 1 January 2004. This American probe, weighing 350 kg, carries a camera and a small mass spectrometer, which will enable it to record the characteristics of the comet, whose nucleus probably does not exceed 2 km in diameter.

The encounter is due to take place 280 million kilometres from the Sun and 400 million kilometres from the Earth. For several minutes, Stardust, protected from impacts by a heavy shield, will approach to within 150 km of Comet Wild 2. During the fly-past, the probe will collect several micro-grammes of particles, a sample of the tail of Comet Wild 2, and will then slowly make its way back to Earth. In January 2006 the precious samples will be returned to Earth.

■ IN THE AUTUMN OF 1997, COMET HALE-BOPP, RECEDING RAPIDLY FROM THE EARTH, WAS STILL CLEARLY VISIBLE AGAINST THE STAR FIELDS OF THE MILKY WAY. IN 1200 YEARS, ITS ORBIT WILL SLOWLY CURVE ROUND AND THE GREAT COMET WILL START ITS LONG RETURN TRIP TOWARDS THE SOLAR SYSTEM'S PLANETS.

In the meantime, another American probe, Contour, will have left our planet to carry out an amazing odyssey across the Solar System. After its departure, in February 2002, Contour will visit, in succession, Comet Encke in 2003, Comet Schwassmann-Wachmann in 2006, and finally, Comet d'Arrest in 2008.

The last American cometary mission of the first decade of the 21st century, Deep Space 4, should be launched in 2003. This small probe is designed to encounter Comet Tempel 1 in 2005, and to land the Champollion module on its surface. It will then return cometary dust and ice to Earth in 2010.

What is undoubtedly the most ambitious cometary mission, however, has been conceived by scientists at the European Space Agency. Their large Rosetta probe will be launched by an Ariane 5 rocket in January 2003, and sent towards Comet Wirtanen, which it will reach in 2011. There, it will go into orbit around the comet's nucleus, and travel with it in its orbit for a year and a half. During 2012, Rosetta will drop a small exploration module onto Comet Wirtanen, which will examine the small icy body and obtain samples, which it will analyze on the spot, thanks to a series of sophisticated instruments. As for Comet Halley, who knows whether a probe will be sent to visit it, when it returns to the inner Solar System, 76 years after its encounter with Giotto?

A rendezvous with the ancient comet will not always be possible, because it will either be swallowed up by an impact with the Sun or Jupiter, or will be ejected from the Solar System, and so will no longer return to grace our skies.

Even when they have disappeared into the solar furnace, or into interstellar space, comets always leave a mark behind them. Their long tails strew tiny particles of mineral dust into space, marking their orbit. Regularly, our planet crosses a cometary orbit and, for several days, passes through a region of space that is full of dust particles. These encounter the atmosphere at a velocity of several thousand kilometres per hour, where they heat up and vaporize, to leave behind them the beautiful, fleeting trace of a meteor – a shooting star.

Neptune:

the great blue sea

■ ON THE FRIGID BOUNDARY OF THE SOLAR SYSTEM,
PLUNGED INTO ETERNAL TWILIGHT, ORBITS NEPTUNE, THE
LAST MAJOR PLANET. TO SCIENTISTS, THIS DEEP BLUE
PLANET QUITE LITERALLY REPRESENTS ANOTHER WORLD.
ALMOST INVISIBLE FROM EARTH – EVEN WITH A LARGE
TELESCOPE – IT FIRST APPEARED IN DETAIL IN AUGUST
1989 THROUGH THE POWERFUL CAMERAS ON
BOARD THE VOYAGER 2 SPACEPROBE.

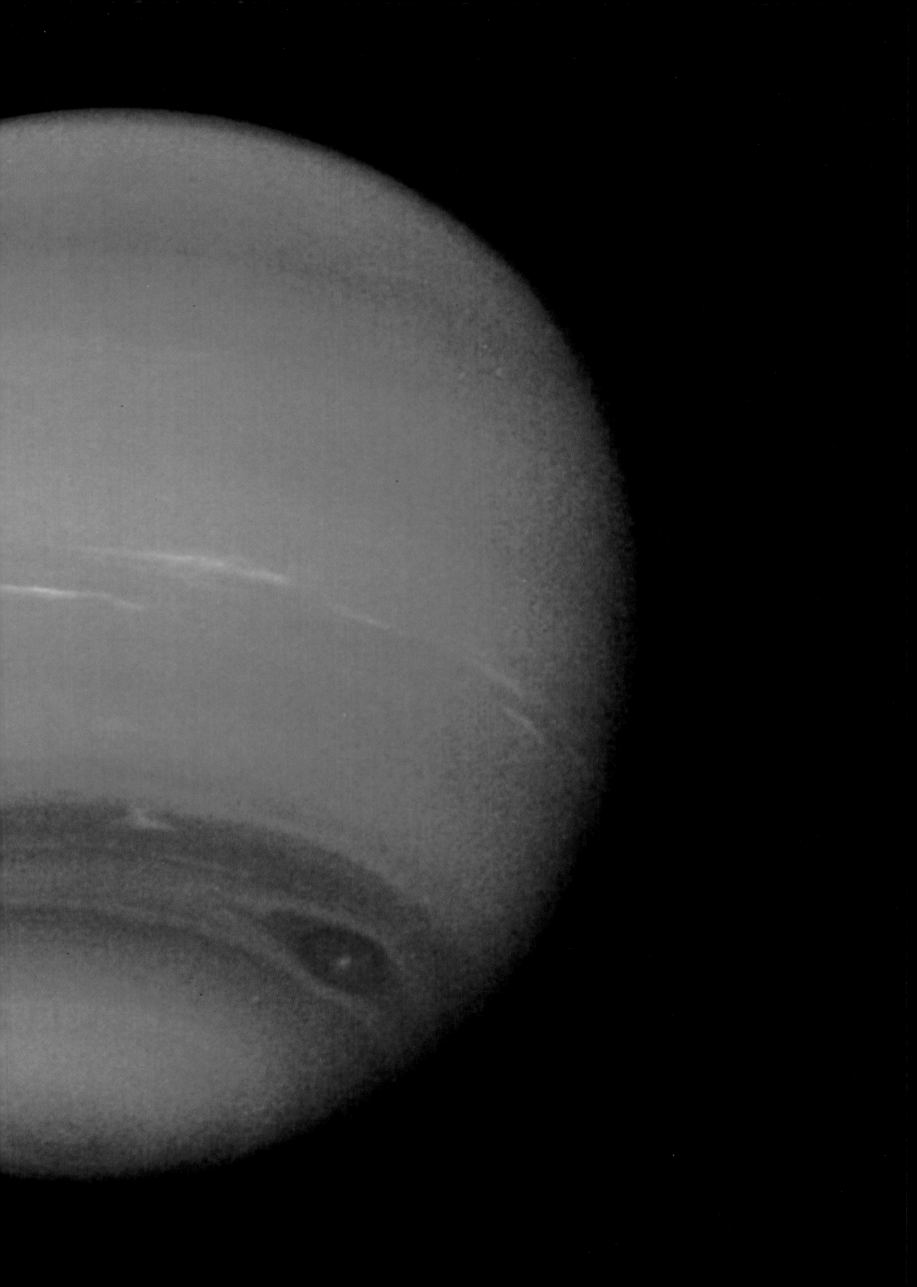

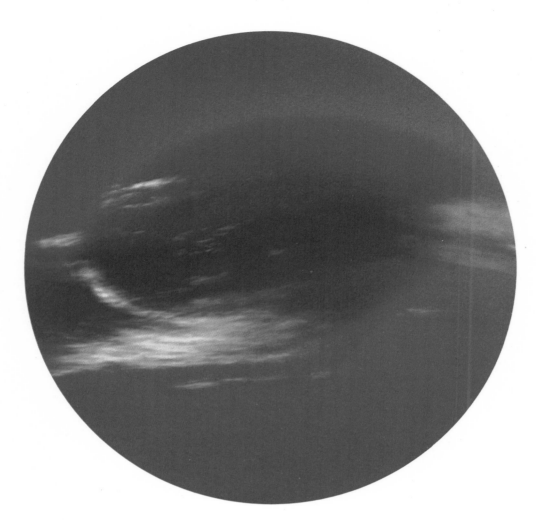

■ NEPTUNE'S ATMOSPHERE IS DISTURBED BY AN ANTICYCLONE OF IMMENSE SIZE: THE GREAT DARK SPOT. CAPABLE OF ENGULFING THE EARTH, THIS GIGANTIC MAELSTROM MAY HAVE PERSISTED FOR SEVERAL CENTURIES IN NEPTUNE'S BLUE ATMOSPHERE.

We are approaching the vicinity of one of the most beautiful planets in the whole Solar System. This body, dark and immense, resembles Uranus, but its azure cloak is stronger in colour and is truly magnificent. It is like flying over the deep blue sea: an infinite expanse, a shadowy abyss 20 000 km deep, smooth and monotonous over tens of millions of square kilometres. Above this breathtaking ocean, a few, startlingly white clouds cast their shadows onto the 'water'. But it is a pure mirage. This vast watery expanse is, in reality, the upper layer of Neptune's atmosphere, a mixture of hydrogen and helium, methane and ammonia. The god of the sea has actually been the sentinel guarding the gates to the Solar System. For twenty years, up to the year 2000, Neptune was farther from us than Pluto, which had been inside the blue giant's orbit for that period. We are in a frigid region of space, plunged into permanent twilight. Neptune, at a distance of 4.5 billion kilometres from the Sun, receives about one thousandth as much energy from our star as the Earth.

The discovery of Neptune is one of the affairs in the history of science that has the most twists and turns in its plot. After William Herschel had discovered Uranus by accident in 1781, the 19th-century mathematicians carried out precise calculations of the orbital elements of the planets in the Solar System. The aim was to establish the motion and characteristics of Uranus as accurately as possible. Laplace and Méchain, and then Bouvard, set out to reveal the gravitational perturbations that Jupiter and Saturn exerted on the new giant planet, which was then thought to be the outermost planet in the Solar System. At the end of the calculations, there remained a small difference between the theoretical and the observed orbits of Uranus. Celestial mechanics was the supreme exact science at that period, so it was probable that neither the astronomers' meticulous observations nor the mathematicians' calculations were marred by errors. So what was the answer? In 1845, on the advice of the great physicist Arago, the young astronomer Urbain Le Verrier, then working at the Ecole Polytechnique, investigated by calculation alone whether a new, hypothetical planet existed, that was assumed to be perturbing Uranus. Le Verrier persevered with the motion of Uranus, and found the orbital elements of the new, unknown planet. He determined the co-ordinates of the body on the celestial sphere, and then,

disdaining to search for himself with a telescope, communicated the position to Johann Galle, an astronomer at the Berlin Observatory, who discovered it immediately, not far from the position predicted by Le Verrier. The result was the glorification of the French astronomer, who was otherwise an odious person, who was later to be the despotic Director of the Paris Observatory for seventeen years. In England, on the other hand, there was consternation for a young researcher, John Couch Adams, who only just missed the discovery because his own calculations, passed to

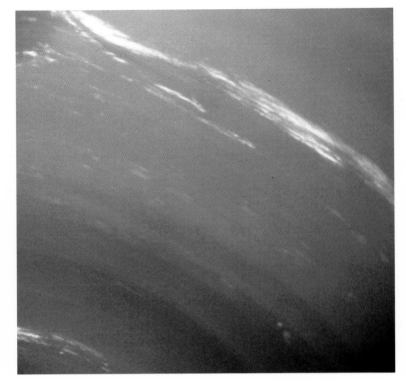

■ LIKE THE ATMOSPHERIC CIRCULATIONS ON JUPITER AND SATURN, NEPTUNE'S OCCURS PARALLEL TO THE EQUATOR. AT THE TOP OF THE CLOUD LAYERS, THE WIND-SPEED REACHES 2200 KM/H, WHICH IS A SOLAR-SYSTEM RECORD. THE TEMPERATURE OF NEPTUNE'S UPPER ATMOSPHERE IS ABOUT −210°C, BUT LOWER DOWN, AT A DEPTH OF ABOUT 3000 KM, IT EXCEEDS 3000°C.

the astronomer George Biddle Airy, were not taken seriously! Who was the true discoverer of Neptune? Adams, Le Verrier, or Galle? Probably none of them. In fact, in 1795, the French astronomer Lalande had recorded a star that moved against the fixed background of the celestial sphere. Unfortunately, the famous astronomer did not pay much attention to his discovery, and attributed it to an observational error. Much more surprising is the fact that the true discovery of Neptune may actually go back as far as – Galileo himself! In 1613, the Italian astronomer had, in fact, recorded in his observation notebooks the movement of Neptune, which was extremely close to Jupiter that year. The great Florentine scholar was within an ace of the discovery, which would have undoubtedly changed a significant portion of the history of astronomy, but he did not pursue the matter.

To this day, and even with the most powerful telescopes, Neptune, which is invisible to the naked eye,

■ NEPTUNE'S UPPER ATMOSPHERE OWES ITS BLUE COLOUR TO THE ABSORPTION OF LIGHT BY METHANE. THIS GAS POSSIBLY REPRESENTS 1 PER CENT OF THE ATMOSPHERE'S CHEMICAL COMPOSITION. THE MAJOR COMPONENTS ARE HYDROGEN AND HELIUM, BUT TRACES OF ETHANE, ACETYLENE, AND AMMONIA WERE ALSO DETECTED BY VOYAGER 2.

reveals practically no detail on its tiny bluish disk, just 2.5" across. Until the 1980s, the 'Neptune' chapter in astronomy books therefore generally amounted to just a single page. All that there was in the way of illustration was a simple photograph, showing a point of light among the stars. As with Uranus, it was Neptune's close encounter with the Voyager 2 spaceprobe in August 1989, that allowed us to acquire most of our knowledge about the last of the gas giants. Voyager 2's encounter with Neptune was literally the result of a technological miracle. The probe was, in fact, originally designed just to visit the planets Jupiter and Saturn. Leaving Earth in 1977, Voyager 2 first crossed Jupiter's system in 1979. Then, as it passed the giant planet, thanks to the gravitational slingshot effect, the probe was sped on its way, at more than 100 000 km/h towards Saturn, which it reached in 1981. At the beginning of the 1980s, by extraordinary luck, the four gas giants happened to be in an exceptional geometrical configuration, which occurs only once every 179 years. The four bodies formed what was, in effect, a sort of spiral, expanding outwards in the same direction as the Solar System's rotation. Neptune was thus ahead of Uranus, itself ahead of Saturn, which was, in turn, ahead of Jupiter. This was a unique chance for NASA's engineers to try to achieve the 'Grand Tour': a visit to all four planets, each stage being accompanied by an acceleration powered by gravity.

So, once the official mission for the two Voyager probes, namely to examine

■ ASTRONOMERS ASSUME THAT NEPTUNE'S STRUCTURE CONSISTS OF THREE PRINCIPAL LAYERS, WHOSE DENSITY INCREASES WITH DEPTH INSIDE THE PLANET. THE TRUE ATMOSPHERE ENDS AT A DEPTH OF ABOUT 6000 KM, BEING REPLACED BY WHAT AMOUNTS TO A MIXTURE OF ICES AND ROCK, PERHAPS WITH SOME METALLIC HYDROGEN. AT THE VERY CENTRE OF THE PLANET THERE MAY BE A ROCKY CORE ABOUT THE SIZE OF THE EARTH.

the systems of Jupiter and Saturn, was finished, Voyager 2 was sent on to Uranus. The probe was, however, beginning to age, partially worn out by the difficult manoeuvres that it had to carry out during the first part of its mission. In addition, the transmission of data, and corrections to the trajectory of a robot that was nearly 3 billion kilometres from the Earth were very delicate operations. Travelling at the speed of light, an order transmitted from Earth took nearly three hours to reach the probe! The Uranus encounter in 1986, was, however, a complete success and persuaded NASA to invest several more hundreds of millions of dollars to attempt a final encounter with distant Neptune.

The engineers had to overcome some extremely difficult technical problems. For example, at Neptune's distance, solar illumination is one thousandth of the strength on Earth. The cameras attached to Voyager 2's telescopes, despite their sensitivity, needed to employ very long exposure times to record the images of the planet and of its satellites. To avoid blurring, the probe needed to follow the proper motion of each body being photographed, which was a computational nightmare. The computers on board the probe were reprogrammed from a distance, data-compression techniques were developed, and the receiving aerials in California, Spain and Australia were specially upgraded. In 1989, twelve years after its departure from

■ NEPTUNE IS ACCOMPANIED BY A SYSTEM OF EIGHT SATELLITES. SIX OF THESE WERE DISCOVERED BY THE VOYAGER 2 SPACEPROBE. EXCEPT FOR TRITON, PROTEUS IS THE LARGEST OF THEM. IRREGULAR IN SHAPE AND HEAVILY CRATERED, PROTEUS IS A BODY, WHOSE CHARACTERISTICS ARE STILL POORLY KNOWN. WITH A DIAMETER OF APPROXIMATELY 400 KM, IT ORBITS NEPTUNE IN SLIGHTLY MORE THAN 24 HOURS, AT A DISTANCE OF 118 000 KM.

Earth, Voyager 2 ended its mission in a blaze of glory as it passed through Neptune's system, sending back images of unsuspected beauty and strangeness.

With a diameter of 49 000 km, Neptune is the smallest of the four gas giants. It is, on the other hand, slightly more massive than Uranus at 100 000 billion billion tonnes, and is, above all, much denser than the other three jovian planets. Its density is 1.73, as against 1.30, 1.34, and 0.70, respectively, for Uranus, Jupiter, and Saturn. Neptune takes slightly more than one hundred and sixty-four years to complete its almost perfectly circular orbit around the Sun.

Like the other three gas giants, Neptune mainly consists of light gases, hydrogen and helium, and a few per cent of methane, which is what gives the planet's upper atmosphere its splendid blue mantle. Unlike Uranus, whose atmospheric phenomena are permanently hidden by haze, Neptune's atmosphere shows considerable detail. Like Jupiter and Saturn, the blue planet is encircled by cloud bands parallel to the equator. These bands are the site of anticyclonic systems, which resemble Jupiter's Great Red Spot, although at a smaller scale.

On Earth, atmospheric disturbances and the cycle of the seasons are intimately linked to the strong radiation from the Sun. On Neptune, however, at the far edge of the Solar System, where the Sun never appears more than a brilliant star in the darkness of the night, solar radiation has practically no effect. Nevertheless, the planet has a surprisingly varied range of atmospheric phenomena. How can these be explained? The answer is that Neptune's atmosphere is undoubtedly driven by the planet's internal radiation. Measurements of infrared radiation from the planet have, in fact, shown that the latter emits three times as much energy as it receives from the Sun. The fringes of Neptune's atmosphere, which are directly exposed to the cold of space, are at an extremely low temperature, less than −220°C. In fact, the atmosphere is primarily influenced by the heat arising in Neptune's high-density, searingly hot core. This sort of inversion of atmospheric phenomena, driven by a form of inner star, affects the planet just as if it were driven by the Sun. The great blue world of Neptune does, in fact, experience storms of indescribable violence, where the winds reach 2200 km/h. The Great Dark Spot, an enormous anticyclone similar to the Great Red Spot on Jupiter, carved out a depression in Neptune's upper atmosphere, and threw cirrus clouds up into the sky, where they dissipated in just a few hours.

The rotation of the planet itself takes 16h9m. Like all the planets in the Solar System, with the exception of Venus and possibly Pluto, Neptune possesses a powerful magnetic field. Magnetic fields arise within the cores of large-sized bodies. Hot, metallic, planetary cores are excellent electrical conductors and their rotation creates a dynamo effect of greater or lesser strength. Neptune's magnetosphere extends several hundreds of thousands of kilometres into space. It traps the flow of charged particles

■ NEPTUNE'S RINGS ARE EXTRAORDINARILY THIN AND WELL-DEFINED. THEY ARE SO NARROW THAT, IN THIS IMAGE, THEY ARE VISIBLE OVER ONLY PART OF THEIR ORBIT, AND APPEAR LIKE LUMINOUS ARCS. IT IS ASSUMED THAT THE CHANGES IN DENSITY IN NEPTUNE'S RINGS ARE CAUSED BY THE PRESENCE OF SMALL SATELLITES WITHIN THEM THAT CREATE GRAVITATIONAL PERTURBATIONS.

■ This impressive photograph of Neptune, obtained by Voyager 2 using special filters, reveals the presence of a thin layer of haze covering the planet's thick atmosphere. This veil is particularly easily seen around the limb of the planet, where it shows as a reddish tint. A few high-level clouds lie not far from the equator.

that is constantly emitted by the Sun. This solar wind, consisting of protons and electrons, concentrates along the lines of magnetic force and, when it reaches the planet's upper atmosphere, ionizes the gases. During the night, when one hemisphere of Neptune is plunged into darkness, it displays magnificent aurorae, like those on Jupiter and the Earth.

Neptune's satellite system is less imposing than those of the other three giant planets, although it includes Triton, the sole satellite, other than Titan, to posses an atmosphere. Six other satellites, discovered by Voyager 2, are no more than large, irregular rocks that are relatively small in size, ranging from 60 to 300 km across. They orbit at distances of between 48 000 km and 118 000 km from the centre of the planet, and their characteristics are very poorly known. The final satellite, Nereid, discovered in 1949, is a very strange body. Its orbit is inclined at 29° to Neptune's equator, and is extremely eccentric. If we exclude most of the comets, its eccentricity is by far the greatest of any body in the Solar System. Nereid takes nearly one year to orbit Neptune. At its closest, Nereid passes less than 1.4 million kilometres from the planet, but it recedes to 9.7 million kilometres at its farthest point! With Triton, which orbits Neptune in a retrograde direction, Nereid poses a difficult problem for scientists. How did a satellite system with such a bizarre confusion of orbits come to be formed?

There is yet another subject for experts in celestial mechanics to investigate and wonder over: the giant planet's

■ THIS DRAMATIC, BACKLIT PHOTOGRAPH OF THE NEPTUNE-TRITON PAIR WAS OBTAINED BY THE VOYAGER 2 SPACEPROBE IN AUGUST 1989 AS IT RECEDED FROM THE PLANET ON ITS ONE-WAY TRIP. SINCE THEN, THE PROBE HAS LEFT THE SOLAR SYSTEM AND, BY EARLY 2002 WAS

system of rings. At first sight these resemble those of Jupiter, Saturn, and Uranus. Neptune's four rings are astonishingly thin. The rings named Le Verrier and Adams are 120 000 km and 100 000 km in diameter, respectively, with widths of just 10 km, and a thickness that is probably less than 100 m! Like the rings of Saturn and Uranus, Neptune's are curiously irregular. There are denser concentrations, known as arcs, around their circumference, whose origin, nature, stability and evolution are quite unknown. Are these structures ephemeral? Will they last for millions of years? We don't know. Researchers suspect, however, that the existence of these arcs is linked to the presence, in the very heart of the

rings, of six satellites. These probably play the role of 'shepherds' in Neptune's system, by confining the dust of the rings to extremely precise orbits. These satellites that orbit within the rings probably also explain the thinness and extraordinarily clear-cut nature of the rings.

Neptune is probably the last true planet in the Solar System, given that the status of the Pluto–Charon pair – which passes inside its orbit once every 200-odd years – is still not clearly established. If we exclude these two small, icy objects, the whole of the Solar System has thus been examined, over a quarter of a century, by a dozen different spaceprobes, of

NEARLY TEN BILLION KILOMETRES FROM EARTH. ITS LONG
VOYAGE ACROSS THE GALAXY HAS HARDLY STARTED, AND
COULD WELL CONTINUE LONG AFTER THE EARTH AND ITS
INHABITANTS HAVE DISAPPEARED.

which Voyager 2 was un-doubtedly the star. As the new millennium begins, the torch has been passed to the Galileo and Cassini spaceprobes, which are looking anew at Jupiter and Saturn, multiplying tenfold our knowledge of the two largest planets in the Solar System. By contrast, however, no missions have been proposed to revisit Uranus and Neptune, which nevertheless still hide numerous secrets.

One day, perhaps before the end of the 21st century, a future generation will witness astronauts disembarking on the surface of some of the fifty-odd satellites that orbit the four gas giants, and which offer an ideal vantage point for their observation. From the surface of Amalthea, Prometheus, Ophelia, or Triton, in the absolute silence of space, the sight of these stormy planets would be astounding.

Triton:

volcanoes of ice

■ Triton is one of the most enigmatic satellites in the Solar System. As large as a planet, possessing a thin atmosphere, geologically very active, it orbits in the opposite direction to most of the planets and satellites. Was Triton born in Neptune's system, or is it perhaps an interloper? To most experts, the body is akin to Pluto and Charon.

■ ONE OF THE MOST CURIOUS GEOLOGICAL FEATURES OF TRITON IS FOUND IN THESE INTERMINGLED GLACIAL RIDGES AND CIRCULAR DEPRESSIONS, BOTH OF UNKNOWN ORIGIN. THIS TYPE OF TERRAIN HAS NOT BEEN OBSERVED ANYWHERE ELSE IN THE SOLAR SYSTEM.

A desert of wrinkled ice, rather like low dunes covered in frost, extends to the horizon. The whole of the landscape is bathed in a crepuscular light, and the ice is uniformly grey and lifeless. In the distance, a few hills that slope towards the south are lit by a soft, faint, bluish light. Towards the east, the icy plain is absolutely flat, covered as far as one can see by perfect polygons, rather like those found on the *salars* of Chile or Bolivia. This planet is one of the strangest in the Solar System. The first astronauts to set foot here, on these polar landscapes, will find themselves disorientated as nowhere else. They will not, however, be put off from exploring this new world. They will feel extraordinarily light; even protected by thick spacesuits that give them an overall mass of nearly 200 kg, they will weigh, here, no more than 12 kg. Will they feel an impression of solitude, a slight bitterness, mingled with anxiety, caused by the great distance from their planet of origin? The Sun, which is the sole reference point common to the whole of our planetary system, will hardly provide any reassurance. Seen from here, it appears incredibly distant. In Triton's sky, our daytime star is minute, and only one thousandth of its brightness on Earth. The

shadows cast on the icy dunes are very faint. The sky is not dominated by the Sun, but by an enormous, sea-blue globe, which slowly rotates, its reflection glancing off the crystal faces of the clear ice. Neptune, as seen from here, has an apparent diameter of 8°, and therefore appears five times as large as the Earth, when seen from the Moon.

The wind has risen. A breeze that would be imperceptible to any possible explorers, armoured against the cold. Triton is the coldest known world in the Solar System. The ice covering its surface reflects between 60 and 90 per cent of the incoming solar radiation back towards the sky. And the exceptionally thin atmospheric veil that covers the planet is unable to retain the heat that is so frugally doled out by our star. The thermometer here reads − 235 °C, an absolute record.

In the distance, something has moved. For just an instant, the ice has flashed and a haze of crystals has erupted from the dreary surface, twinkling in the icy light from the Sun. Beyond the horizon, an enormous geyser now rises into the star-studded sky. Arising in the depths of the planet, where the higher temperature first liquidizes and then vaporizes it, the nitrogen, heated to nearly − 100 °C erupts at several hundreds

■ THIS ALMOST SMOOTH ICY
PLAIN, BROKEN BY A SINGLE
IMPACT CRATER, IS GEOLOGICALLY
VERY YOUNG AND ONE OF THE
MOST INTERESTING REGIONS ON
TRITON. THE ORIGIN OF THE
LARGE EXPANSE MIGHT BE
EXPLAINED BY SUBTERRANEAN
WATER THAT HAS SPREAD
OUT ON THE SURFACE AND
FROZEN WHEN EXPOSED
TO THE FRIGID COLD OF
SPACE.

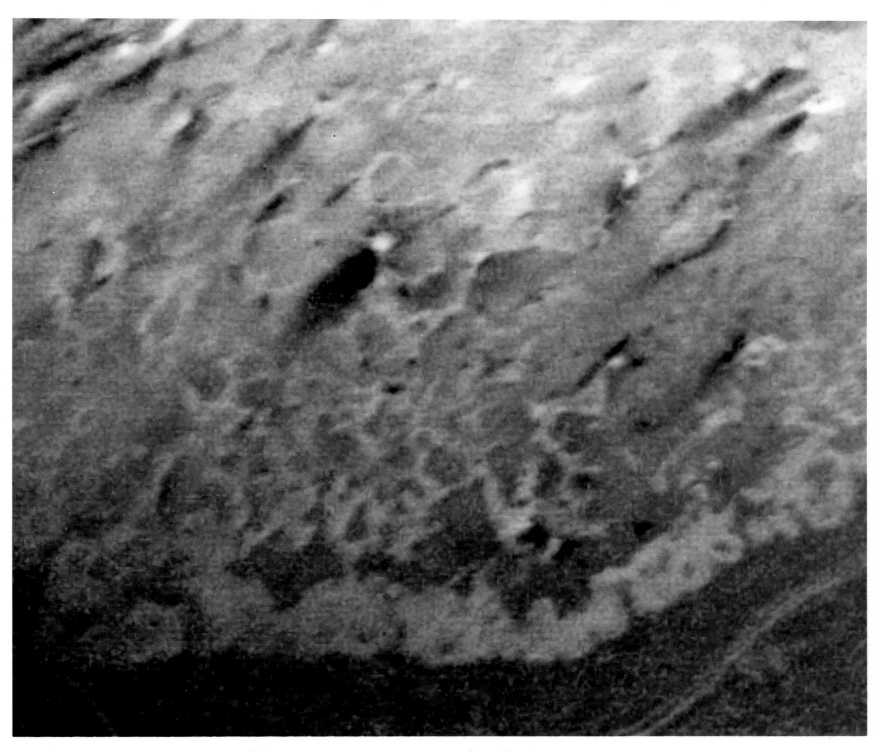

■ A HIGH-RESOLUTION IMAGE OF THE SURFACE OF TRITON. THE SMALLEST DETAILS VISIBLE HERE ARE 4 KM ACROSS. SMALL WHITE SPOTS ARE ACCOMPANIED BY DARK, ELONGATED PLUMES. THESE ARE PROBABLY GEYSERS AND THEIR DEPOSITS, SPREAD OUT BY THE PREVAILING WIND ON THE SATELLITE. ONLY THE EARTH AND IO CURRENTLY EXHIBIT SIMILAR VOLCANIC ACTIVITY.

of kilometres per hour. The geyser rises to a height of slightly more than 8000 m; there the gaseous plume slowly spreads out under the influence of the prevailing wind, cools, condenses into crystals which fall back as snow onto the frigid surface of Triton, sometimes as much as 100 km from the site of the eruption.

Triton is by far the largest of the eight known satellites of Neptune. It measures more than 2700 km in diameter and its mass is 20 billion billion tonnes. This places it fifteenth in rank in the Solar System, after Europa, which is twice as massive, and just ahead of Pluto, whose mass is scarcely 13 billion billion tonnes. Its density of 2.0 is practically equal to that of Ganymede, Titan, and Pluto (1.93, 1.9, and 2.1, respectively). Triton takes 5 days 21 hours to complete its perfectly circular orbit around Neptune at a distance of 354 000 km from the centre of the planet. Gravitationally locked to Neptune, it permanently turns the same hemisphere towards the planet, because its periods of revolution and rotation are identical. It may be noted that the distance between Neptune and Triton is practically the same as that between the Earth and the Moon. Nevertheless, the latter takes about one month to complete an orbit around the Earth. Why does Triton orbit its planet about five times as fast? The answer is found in the mass ratios between the two pairs of bodies. The Earth, which drags the Moon around with it, is eighty-one times as massive as its satellite. Neptune, which controls Triton's orbital revolution, is five thousand times as massive! Triton's orbital velocity is much greater than that of the Moon, so it thus escapes the immense, and otherwise fatal attraction of the giant planet. After the Moon, and the atypical case of Charon, Pluto's satellite, Triton holds the record for its mass relative to its parent planet.

Many common features link Triton, Titan, and the Pluto–Charon pair: their great distance from the Sun, their sizes, their comparable masses and densities, and, finally, their atmospheres. Triton does, in fact, have a very thin atmos-

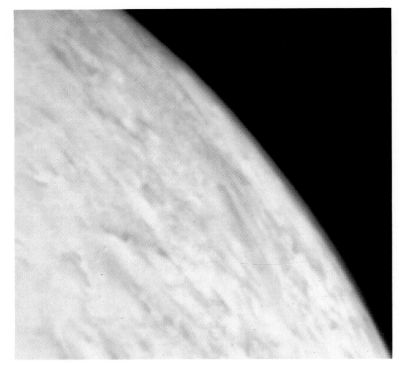

■ THIS IMAGE FROM VOYAGER 2 CLEARLY ILLUSTRATES THE DIVERSITY OF THE GEOLOGY ON TRITON. IN THE CENTRE, TWO CANYONS EXCAVATED IN THE ICE, RUN ALONGSIDE AN AREA OF UNEVEN AND COMPLEX RELIEF. ON THE LEFT, GEYSERS PUNCTUATE A PLAIN MOTTLED BY FROST. VERY RECENT, THIS REGION SHOWS JUST ONE IMPACT CRATER, AT POINT WHERE THE TWO CANYONS DIVIDE.

phere, probably identical to that of Pluto, and whose composition resembles that of Titan. Like the latter, the atmosphere of Neptune's large satellite essentially consists of nitrogen and methane. On Titan, closer to the Sun, the temperature averages −180 °C, which, although extremely low, enables nitrogen to exist as a solid, liquid, or gas. On Triton, by contrast, the atmospheric nitrogen is probably almost completely frozen out onto the surface. In fact, the atmospheric pressure here is the lowest that we have measured anywhere in the Solar System: 0.000 015 bar. The density of Triton's atmosphere is thus 65 000 times less than that measured at sea level on Earth. Although extremely tenuous it is nevertheless detectable on the photographs taken by Voyager 2, as a very thin veil uniformly covering the planet at an altitude of about 5000 m. Like the haze on Titan, that on Triton consists of methane and ethane. Unlike the Earth's atmosphere, where the temperature of the lowest layer decreases evenly with altitude, the temperature on Triton attains its maximum value – almost −100 °C, in this thin cloud layer.

Triton permanently orbits within Neptune's magnetosphere. Because it is surrounded by an atmosphere, the charged particles trapped by the giant planet's powerful magnetic field reach the surface, through the thin atmosphere, where the nitrogen atoms have time to be ionized. So, not content with observing ice volcanoes on

Triton that are capable of expelling plumes of nitrogen as high as Daulaghiri, a Sun that looks like a brilliant point of light, and the sea-blue globe of a giant planet hanging motionless in the sky, future explorers of the Solar System will also have the rare privilege of admiring the blaze of magnificent aurorae on Triton.

Neptune's large satellite was once one of the most active worlds in the whole Solar System. Although its icy volcanism still indicates significant internal activity, its past is revealed as being incomparably richer. Like that of most of the satellites of Jupiter, Saturn, and Uranus, Triton's crust is partially covered in impact craters. Yet vast areas of the surface are practically devoid of them, and instead exhibit curious icy festoons and ridges, whose origins are still mysterious. In one area, the surface seems to be cracked, and strange, irregular, dark patches are surrounded by lighter collars, probably consisting of clear ice. In another, an ancient basin appears to have been buried beneath a flow of smooth, evenly grey, ice. Like Titan, Triton probably has a dense, rocky core, surrounded by concentric layers of ices that are lighter towards the surface. The astonishing topography (which has been formed by intense tectonic movements) that is found on this small world can only be explained by a turbulent past.

We can try, very partially, to reconstruct the history of Triton by studying the utterly exceptional orbit that it

■ THIS EXCEPTIONAL IMAGE SHOWS TRITON'S EXTREMELY THIN ATMOSPHERE, ABOVE THE ICY SURFACE OF THE WORLD. THIS TENUOUS ENVELOPE APPEARS BLUISH. THE TEMPERATURE OF THE SURFACE OF TRITON DOES NOT EXCEED −235 °C: THE SOLAR SYSTEM'S RECORD FOR A LOW TEMPERATURE. THE ATMOSPHERIC PRESSURE IS 65 000 TIMES LESS THAN THAT AT THE SURFACE OF THE EARTH.

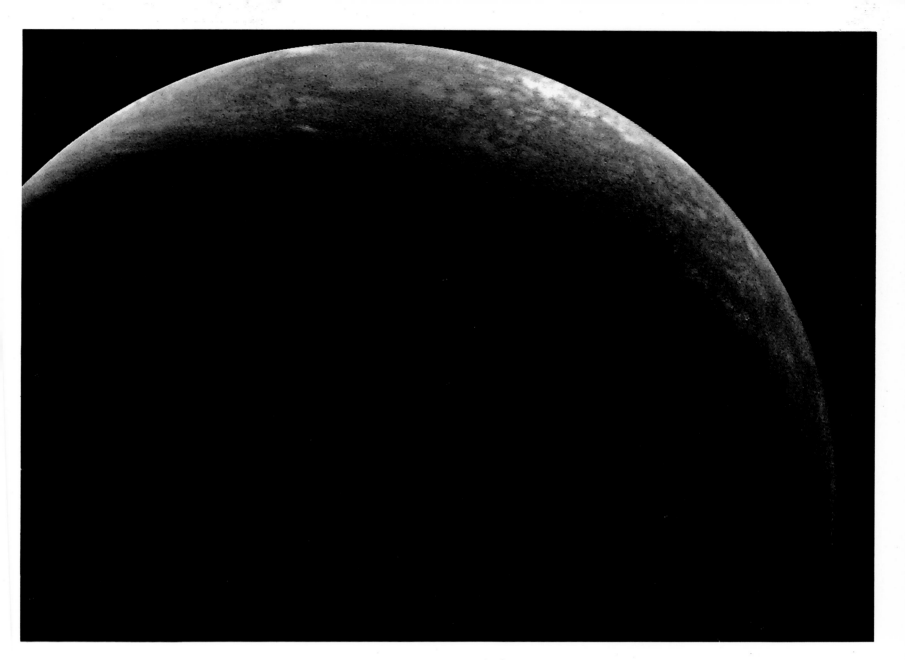

■ Triton's fate is as mysterious as its origin. Certain astronomers predict that after slowing in its orbit around Neptune, the large satellite will slowly approach the planet. In less than 100 million years, Triton may even drop below the Roche Limit, and be fractured by the giant planet's gravitational forces.

follows around Neptune. We have seen that this is circular and that Triton, like most satellites, permanently turns one hemisphere towards its planet. However, Triton's orbit is extremely surprising. It is inclined at 157° to Neptune's equator – whereas satellite orbits in general are strictly confined to the same plane as their planets' equators – and, above all, Triton's orbit is retrograde. This means that, seen from an imaginary point lying above the north pole of the Sun, Triton orbits Neptune in a clockwise direction, opposite to all the planets and all the major satellites in the Solar System! Of the eighty satellites currently known in our planetary system, just ten (Ananke, Carme, Pasiphae, Sinope, Phoebe, Triton, and four small satellites of Saturn) orbit in the opposite direction. The first four of these are no more than rocky blocks some thirty-odd kilometres across that follow immense orbits, highly inclined to Jupiter's orbital plane. These are almost certainly captured asteroids. Phoebe, with a diameter of 200 km, and the four other, small retrograde satellites of Saturn have also probably been captured by the planet.

■ Lying as it does, like them, at the edge of the Solar System, Triton possibly has much in common with Pluto and Charon. Like Triton, these small icy worlds are surrounded by extremely tenuous atmospheres.

But what are we to make of Triton? Its size, its density, and its composition prevent it from being classed as an asteroid. Most experts consider that it – together with Pluto and Charon, which we shall discuss in the last chapter – is a miraculous survivor from the time the Solar System originated. Triton, formed at a great distance from the Sun, like the four gas giants, just escaped a collision with Neptune, and was captured in a retrograde, highly inclined, and highly elliptical orbit by the blue giant. Over time, and with the gravitational friction produced by Neptune, the orbit of Triton eventually became circular. Over the course of hundreds of millions – or perhaps even billions? – of years that this process lasted, Neptune's gravitational tides may have partially melted Triton's interior, and caused the geological upheavals on its surface that may still be observed today.

Pluto and Charon:

planets in limbo

■ PLUTO, THE OUTERMOST PLANET IN THE SOLAR SYSTEM, AND CHARON FORM A STRANGE PAIR, WITH AN UNUSUAL ORBIT, AND WITH ENIGMATIC ORIGINS. TO ADD TO THE MYSTERY, PLUTO AND CHARON ARE THE ONLY PLANETS NEVER TO HAVE BEEN VISITED BY ANY SPACEPROBES. THE BEST IMAGE OF THE DOUBLE PLANET WAS OBTAINED BY THE HUBBLE SPACE TELESCOPE.

■ PLUTO IS THE LAST PLANET TO BE DISCOVERED, AFTER A SYSTEMATIC SEARCH CARRIED OUT AT FLAGSTAFF, IN ARIZONA, BY THE AMERICAN ASTRONOMER CLYDE TOMBAUGH. THIS IS THE DISCOVERY PHOTOGRAPH, TAKEN ON 29 JANUARY 1930.

The ground is covered in carbon-dioxide ice as hard as steel, together with a few black rocks. As far as the eye can see the surrounding landscape consists of low, pale hills, bathed in a feeble, crepuscular light. Everywhere there are merely dismal variations on a single grey, lifeless, dispiriting theme. The heavens are as black as velvet, pricked by stars, and crossed by the silvery arc of the Milky Way, which is an overwhelming presence, sparkling with stars. Vega, Deneb, and Altair, brilliant diamonds in the sky, form a large triangle in the summer sky. The thermometer shows $-220\,^{\circ}$C, and it is early morning on the planet Pluto, but the temperature will not rise during the day. Above the western horizon, suspended in the sky and strangely immobile, the enormous, frigid globe of Charon glimmers with a pallid light. The dark patches that mark its surface do vaguely resemble the terrifying face of the ferryman to the dead.

To the east, among the multitude of stars in Taurus, one extraordinarily bright star attracts the eye. It is its light that sparkles from the splinters of ice, as brittle as glass, that litter the ground. It also manages to cast a faint shadow behind the strangely shaped rocks and the large columns of ice that project

from the barren plain. The Sun. Seen from here, however, it is unrecognizable. To the naked eye it is no more than a star, a perfect point of light of unchanging, eerie brightness, that gives off no heat, 1500 times fainter than the Sun on Earth, and scarcely 200 times as bright as the Full Moon.

Human being like to establish their territory, to define limits, and to mark out their universe with more-or-less symbolical boundaries. The Solar System's frontier guard is Pluto, whose orbit encircles those of all the other known planets, at a distance of about 6 billion kilometres from the Sun.

The most distant of the planets is also the most enigmatic. We know nothing about its origin, nothing about its internal composition, and nothing about its morphology. No human-built robot has ever visited it. Even more: no space mission has been proposed to visit it in the near future. There is just one hope for astronomers, the Pluto-Kuiper-Express mission, which American scientists are hoping to get NASA to accept. This would be a small, automatic spaceprobe, which could leave Earth in 2004 to reach Pluto in 2014. The current position of Pluto and Charon on their orbit is exceptionally favourable, and makes them particularly easy to reach. If,

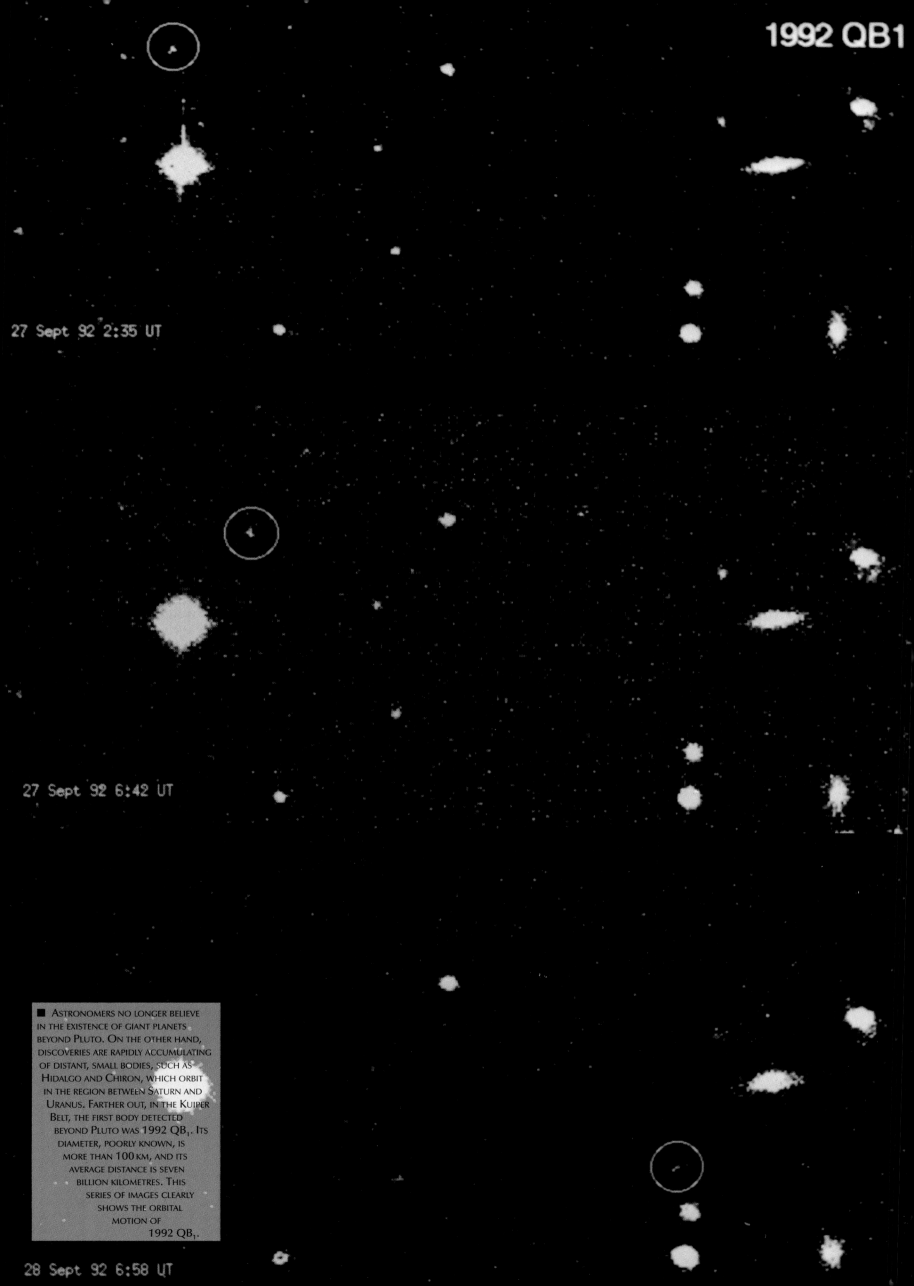

1992 QB1

27 Sept 92 2:35 UT

27 Sept 92 6:42 UT

■ ASTRONOMERS NO LONGER BELIEVE
IN THE EXISTENCE OF GIANT PLANETS
BEYOND PLUTO. ON THE OTHER HAND,
DISCOVERIES ARE RAPIDLY ACCUMULATING
OF DISTANT, SMALL BODIES, SUCH AS
HIDALGO AND CHIRON, WHICH ORBIT
IN THE REGION BETWEEN SATURN AND
URANUS. FARTHER OUT, IN THE KUIPER
BELT, THE FIRST BODY DETECTED
BEYOND PLUTO WAS 1992 QB_1. ITS
DIAMETER, POORLY KNOWN, IS
MORE THAN 100 KM, AND ITS
AVERAGE DISTANCE IS SEVEN
BILLION KILOMETRES. THIS
SERIES OF IMAGES CLEARLY
SHOWS THE ORBITAL
MOTION OF
1992 QB_1.

28 Sept 92 6:58 UT

unfortunately, this mission does not see the light of day, it will not be possible to visit Pluto and Charon for some decades, because their distance from the Sun and their inclination relative to the mean plane of planetary orbits are both increasing year by year.

So Pluto risks remaining for a long time no more than a simple luminous disk in the sky, and in the years to come only giant telescopes will offer any chance of discovering more about the Solar System's most mysterious body. Nevertheless, astronomers on Planet Earth have already managed to collect an impressive amount of information from astrometric data, and from photometric and spectroscopic measurements of Pluto. The planet, discovered in 1930 by Clyde Tombaugh, has since become better known. Its day lasts almost a terrestrial week, and the year, 248 of our years. Pluto measures 2400 km in diameter. Its density of just 2.1 means that it is a light body, having a thick mantle of ices. Pluto's mass is approximately one five-hundredth of that of the Earth. The object is therefore really tiny: much smaller than Mercury, Titan, Triton, the Moon, Io, Europa, Ganymede, and Callisto.

As for Charon, its diameter is about 1200 km, and its mass is roughly one tenth of that of Pluto. Its density, according to the latest observations made with the Hubble Space Telescope, in 1993, is just 1.4. In size, it therefore resembles some of the smaller satellites of Saturn, such as Rhea, Tethys, Iapetus, and Dione. Yet compared with Pluto, Charon is an enormous satellite. Rather than speaking of a planet and its satellite, Pluto and Charon should be considered as a double planet. In addition, the two bodies are almost literally touching one another. According to measurements made by the Hubble Space Telescope in 1991, their centres are 19 000 km apart, so their surfaces are within 17 000 km of one another! For comparison, the Earth-Moon distance is about 380 000 km. Charon orbits Pluto in 6.39 days, and this period of revolution is equal to the planet's period of rotation. In other words, the two bodies act as if they are joined by an invisible link, and permanently face one another.

The situation found at the surface of Pluto is unique in the whole Solar System. From one whole hemisphere of Pluto, Charon is never visible. From any other point on the planet the satellite remains fixed in the sky, and never moves at all. Over the course of the

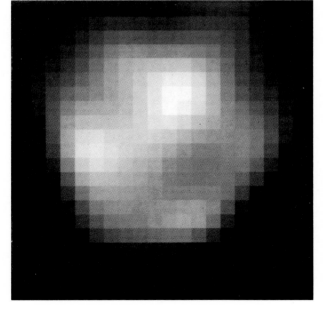

■ THIS IS AN EXCEPTIONAL IMAGE OBTAINED WITH THE EUROPEAN CAMERA ON THE HUBBLE SPACE TELESCOPE. IT SHOWS, FOR THE FIRST TIME, PATCHES ON THE SURFACE OF PLUTO. PLANETOLOGISTS BELIEVE THAT THE SMALL PLANET CLOSELY RESEMBLES TRITON, NEPTUNE'S SATELLITE. THIS IMAGE, GREATLY ENLARGED, SHOWS THE HUBBLE CAMERA'S INDIVIDUAL PICTURE ELEMENTS (PIXELS).

weeks, its phase changes, like that of the Moon, and the sky, the planets and the Sun, slowly move behind this enormous pale globe, which is eight times as large as the Moon seen from Earth. Let us go to Pluto's equatorial region. There, Charon remains eternally at the zenith, whereas at the poles its disturbing pallid globe – half hidden behind the frigid hills that form the horizon – never rises.

Despite the intense cold that seems to have held the planet permanently in its grip, Pluto is currently enjoying milder conditions than it experiences throughout most of the two-and-a-half centuries that it takes to circle the Sun. At the end of the 20th century it was, in fact, summer on the planet. Its extremely eccentric orbit means that its distance from the Sun varies between 4.4 and 7.5 billion kilometres. From early 1979 until early 1999, Pluto, at a distance of about 4.5 billion kilometres was closer to the Earth and the Sun than Neptune! This, as we have seen, offers a unique occasion to launch an exploratory mission to the double planet. The unusual amount of radiation from the Sun appears to have been sufficient to produce seasons on Pluto and Charon. A highly tenuous atmosphere of nitrogen, carbon dioxide, and methane currently surrounds the two bodies. When autumn arrives in the first decades of the 21st century, it is probable that the cold will cause this thin, frigid envelope to freeze onto the surfaces of Pluto and Charon. This layer of frost, deposited on the glacial expanses of the outermost planets in the Solar System, will persist for nearly two centuries, before it sublimes – next summer.

And beyond that? What is there beyond the last planets in the Solar System? The apparently vast distance that separates us from Pluto – forty times the Earth–Sun distance – is infinitesimal on a cosmic scale. A spaceprobe, racing along at 100 000 km/h would require ten years or so to cover the distance that separates us from Pluto, a distance that light covers in less than six hours! The distance of the closest star to the Sun, Proxima Centauri, is, for comparison 40 000 billion kilometres, an abyss that light takes slightly more than four years to cross. The Solar System's true frontier is extremely difficult to define, but lies somewhere between our star and Proxima Centauri, at a distance of one or two light-years. It is there, at more than 10 000 billion kilometres from the Sun, but still within its gravitational influence, that we find the Oort Cloud, a

■ In February 1994, after being repaired in orbit, the Hubble Space Telescope aimed the European high-resolution camera at Pluto and Charon, allowing a new, very accurate, measurement of their diameters, which are 2320 km and 1270 km, respectively. The Hubble observations also showed that Pluto's atmosphere consists of a very slight haze.

sort of quasi-spherical shell consisting of some 1000 billion comets, which, all together, amount to about the mass of the Earth. Is it possible, however, that other, still unknown planets exist within the confines of the Solar System? People have thought so for many years, and some astronomers have spent their lives searching for them. But massive bodies, like Jupiter or Uranus, would already have been spotted, either by the systematic observation of the sky, or by calculation, through the perturbations that they would have exerted on the Solar System within their orbits. We now believe that the known planets, from Mercury to Neptune, have swept up and accumulated practically all the material available in the forming protoplanetary disk, some 4.6 billion years ago. We are left with the comets, which either crashed into the primordial planets or were ejected out to the Oort Cloud. But we are also left with a myriad bodies that were utterly unknown some ten years ago and which, as we now know, populate the Solar System between Pluto and the comets in the Oort Cloud. There is in fact a whole new 'world' that astronomers are starting to explore with the beginning of the 21st century. The first of these objects was discovered in 1992 by Jane Luu and David Jewitt, two particularly patient, American scientists. Patient, because for five years they had been searching the sky for proof of the existence of these icy bodies, which had been predicted by Gerard Kuiper in 1950, but which were far too small and distant to be detected by telescopes at that time. The body, 1992 QB_1, is a sort of

asteroid covered in ice, some one hundred kilometres in diameter, whose average distance from the Sun is more than 7 billion kilometres. It turned out quite quickly, as observations of these icy worlds accumulated, that 1992 QB_1 was merely the first of a new class of bodies, now called 'transneptunian objects' by the scientists. At the beginning of 2002, some 500-odd of these icy mini-planets are known, which formed far from the Sun some 4.6 billion years ago, and which, under the influence of gravitational perturbations by the giant planets, have eventually stabilized in a sort of diffuse torus, known as the Kuiper Belt, that lies between 30 and 50 astronomical units from the Sun, i.e., between 4.5 and 7.5 billion kilometres. There may, according to the experts, be 100 000 transneptunian objects larger than 100 km across, orbiting in this region! Statistically, scientists estimate that there may be a total of more than one billion of these tiny worlds, including all those more than 5 km across. So the overall mass of the transneptunian objects may exceed 100 billion billion tonnes, ten times as much as all the asteroids orbiting between Mars and Jupiter!

Two large transneptunians were discovered in 1996: 1996 TL_{66} and 1996 TO_{66}, which are approximately 500 km and more than 600 km in diameter, respectively. In 2000 and 2001 even bigger objects were discovered. Varuna (2000 WR_{106}) and 2001 KX_{76} – as yet unnamed – are about 900 km across (or perhaps even larger), and appear both in size and chemical composition to resemble the planetary oddities of Pluto, Charon, and

■ As the 21st century begins, we know more than 500 transneptunian objects. Astronomers found 1994 TG_2 at about 6.3 billion kilometres from the Sun, that is, beyond the orbits of Neptune and Pluto. This object, 1994 TG_2, probably measures about 100 km across.

17 May 1993, 0h46 UT

18 May 1993, 0h24 UT

■ OBSERVED FOR THE FIRST TIME IN MARCH 1993, THIS TRANSNEPTUNIAN OBJECT, LYING WELL BEYOND THE ORBITS OF NEPTUNE AND PLUTO, PROBABLY MEASURES SEVERAL HUNDRED KILOMETRES IN DIAMETER. IN THESE IMAGES, THE NEW MINOR PLANET IS SEEN MOVING IN FRONT OF A BACKGROUND OF STARS AND DISTANT GALAXIES.

Triton. These three mini-planets are, in fact, so similar that they must have a common origin. They have similar sizes, masses, and compositions. According to recent studies, astronomers believe that Triton, Pluto, and Charon were formed at the same time as the gas giants, when the Solar System originated, and are the sole survivors of a gigantic game of cosmic billiards. Triton was captured by Neptune. Pluto was placed in an orbit that crosses inside that of Neptune, but protected against colliding with the gas giant by a particular gravitational effect, a resonance, which prevents the two bodies from approaching one another. Finally, Charon, a satellite that is quite abnormal, both in size and distance from its parent planet, is undoubtedly the result of a collision with, and subsequent capture by, Pluto. The violence of this collision might explain Pluto's 'distorted' orbit around the Sun. When the crash occurred, Charon would have caused most of the ice originally present on Pluto to be ejected into space. Subsequently, the satellite would have accreted some of the ice lost by the planet. This scenario would explain the difference in density of the two bodies. While this would place Triton, Pluto, and Charon in Neptune's family, thousands of similar bodies were blindly orbiting the outer edge of the Solar System. Many of these must have crashed onto Jupiter, Saturn, Uranus, and Neptune. The others, only just avoiding collision, were violently ejected from the inner Solar System through gravitational slingshot effects caused by the four gas giants, and are now calmly orbiting within the Kuiper Belt. Finally, some of the small planets were probably ejected from the Solar System for good. Some of them have probably been wandering for 3 or 4 billion years in the vast depths of the Galaxy, while others may even have found refuge around other stars…

POSITIONS AND MOTIONS
IN THE SOLAR SYSTEM

Before studying the motions of bodies in the Solar System, and the laws that govern them, it is as well to be familiar with the proportions of the system, as seen from outside. It is dominated by a star, the Sun, which emits a vast amount of energy, a large portion in the visible region. At first sight, one would not notice anything other than empty space and darkness around this brilliant body. Our eyes are dazzled by the Sun, and the planets, which reflect only a tiny fraction of its light, are very difficult to detect. Their brightness is about one billionth of the Sun's. Similarly, their sizes and masses are insignificant when compared with those of the Sun. The mass of all the planets in the Solar System represents 1/1000 of a solar mass, and the mass of the Earth 1/300 000. But this impression of empty space prevailing around the Sun is confirmed if we compare the distances of the planets from the Sun relative to their actual dimensions. If the Sun is represented by a tennis ball, say 10 cm across, the Earth would be represented by a tiny ball, 1 mm in diameter, at a distance of 11 metres. Saturn would be a ball less than 1 cm across at 100 m, and Pluto, a minute sphere just 0.3 mm in diameter, would then be 400 m distant. On this scale, the closest star to the Sun would be at a distance of some 2000 km ... This distance may be compared with the 500 m (on the same scale) which modern spaceprobes have reached.

KEPLER'S LAWS

Everyone has known since the time of Nicholas Copernicus that the Earth and all the planets orbit the Sun. In an idealized Solar System, all the planets would orbit in a single plane, on perfectly circular orbits; each planet's plane of rotation (its equatorial plane) would be identical to its plane of revolution (its orbit). The same would apply to all the other bodies: satellites, asteroids, and comets. The actual situation regarding planetary motion is rather different, and might be compared with the apparent clarity of motion of the hands of a clock and the complexity of the underlying clockwork.

To simplify the following explanation, we restrict ourselves to the nine major planets and their satellites, which represent 99.9 per cent of the Solar System's mass, excluding the Sun itself.

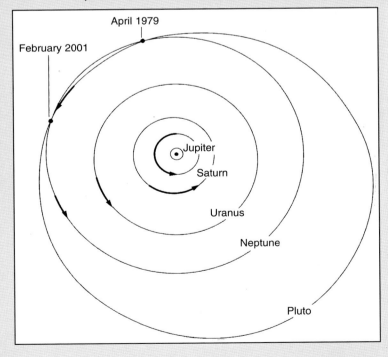

■ APART FROM THE ORBIT OF MERCURY AND PLUTO, THE PLANETARY ORBITS ARE PRACTICALLY CIRCULAR. THIS DIAGRAM SHOWS THE PATHS OF THE FIVE OUTER PLANETS, BETWEEN THE YEARS 1993 AND 2000. THE ORBIT OF PLUTO INTERSECTED THAT OF NEPTUNE IN 1979. PLUTO BECAME THE SOLAR SYSTEM'S OUTERMOST PLANET AGAIN IN FEBRUARY 2001.

One of the tricky problems that confronts observers, when studying the motion of the planets, is to discriminate between the apparent and the true paths of planets on the two-dimensional celestial sphere. Discovering the laws of a system in motion, when the observers themselves are moving within the system, is not a simple task. Even now that the conquest of space has begun, we still need to make a real intellectual effort to convince ourselves, when we admire the apparent path of the Sun across the sky, that it is the Earth that is rotating on its axis. It was this limitation of the human mind, which prevents it from grasping clearly the geometry of space and from changing reference frames, that meant that the Ptolemaic system prevailed until the time of Copernicus, despite the heliocentric system having been anticipated in antiquity.

It is to Johannes Kepler that we owe the first explanation of the apparent movements of the planets on the celestial sphere, and its application to the true motion of the planets in three-dimensional space. Kepler formulated three fundamental laws governing planetary motions, laws that equally apply – as Kepler could not have known – to almost every motion of bodies throughout the universe.

The first of Kepler's laws states that the orbit of a planet is not perfectly circular, but elliptical. The difference between this ellipse and a circle is measured by its eccentricity, which may vary between 0 and 1. This value is often

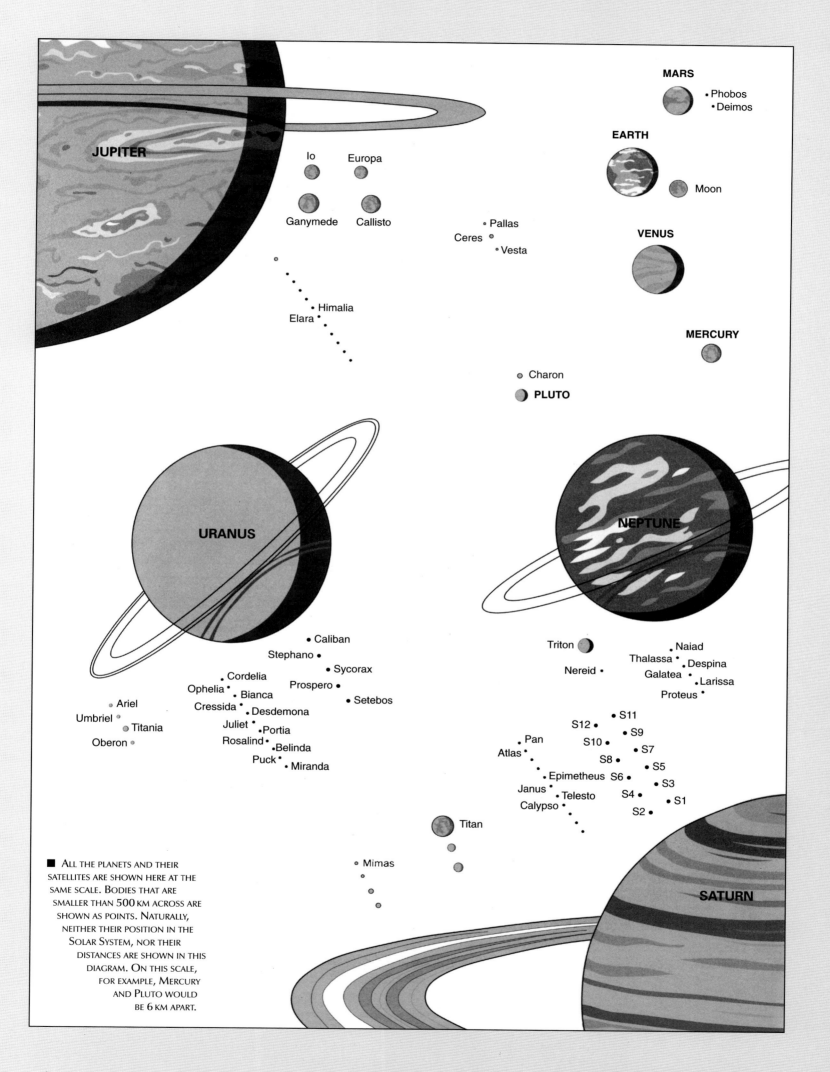

JUPITER

MARS
• Phobos
• Deimos

EARTH

Moon

Io   Europa

Ganymede   Callisto

• Pallas
Ceres ○
• Vesta

VENUS

• Himalia
Elara •

MERCURY

○ Charon

◐ PLUTO

URANUS

NEPTUNE

Triton ◐
• Naiad
• Caliban
Stephano •
• Sycorax
Prospero •
• Cordelia
Ophelia •
• Bianca
Cressida •
• Desdemona
Juliet •
• Portia
Rosalind •
• Belinda
Puck •
• Miranda
• Setebos

Thalassa •
Nereid •
• Despina
Galatea •
• Larissa
Proteus •

• Ariel
Umbriel ○
○ Titania
Oberon ○

S12 •
• S11
• S9
Pan •
S10 •
• S7
Atlas •
S8 •
• S5
• Epimetheus S6 •
• S3
Janus •
• Telesto S4 •
• S1
Calypso •
S2 •

Titan

• Mimas

■ ALL THE PLANETS AND THEIR
SATELLITES ARE SHOWN HERE AT THE
SAME SCALE. BODIES THAT ARE
SMALLER THAN 500 KM ACROSS ARE
SHOWN AS POINTS. NATURALLY,
NEITHER THEIR POSITION IN THE
SOLAR SYSTEM, NOR THEIR
DISTANCES ARE SHOWN IN THIS
DIAGRAM. ON THIS SCALE,
FOR EXAMPLE, MERCURY
AND PLUTO WOULD
BE 6 KM APART.

SATURN

extremely low, as in the case of Venus, whose distance from the Sun varies only between 107 and 109 million kilometres. The eccentricity may also take extreme values, as is the case with many asteroids and comets, and even with Nereid, one of Neptune's satellites.

First Law: *A planet describes an ellipse around the Sun, which lies at one of the two foci.*

Kepler's second law introduces the notion of the velocity of the bodies that orbit the Sun. It shows that the velocity of a planet in an eccentric orbit varies constantly, reaching a maximum value near the Sun.

Second Law: *The radius vector (the line) joining a planet to the Sun sweeps out equal areas in equal times.*

The third law is the most interesting. It enables one to calculate the distance of any planet, once its orbital period is known, together with the distance between the Earth and the Sun.

Third Law: *The squares of the orbital periods are proportional to the cubes of the major axes of the orbits.*

But although these three laws explain remarkably well how the planets move, they do not explain why.

FROM NEWTON TO EINSTEIN

The answer came from Sir Isaac Newton, the distinguished British mathematician and physicist, to whom we also owe, among other inventions, that of the reflecting telescope. Newton introduced the mechanical concepts of force and mass into the motion of the planets. For example, he understood that the force that, on Earth, causes an apple to fall towards the centre of our planet is the same as the force that, in space, forces the Moon to orbit the Earth. This is the

force of gravity, or gravitational attraction, which at the surface of a planet, appears in the form of weight. In 1687, he published his famous law of gravitation, which may be expressed, in shortened form, as: *'Two bodies attract one another with a force whose strength is directly proportional to the product of their masses, and inversely proportional to the square of their distance.'*

In this way, Newton created a whole new science, that of celestial mechanics, and endowed it with an extraordinary scope, because he postulated that the law of gravitation is – universal.

Since the beginning of the 20th century, the law of universal gravitation has been incorporated in the theory of general relativity, which gave the universe a new, and very strange, aspect. According to general relativity, the planets that orbit the Sun are no longer simply subject to forces that keep them in their orbits in space that is, by definition, empty and passive.

Space – or rather four-dimensional space-time in Einstein's theory – is active.

It is distorted by the presence of a material body, just as rays of light that pass near a similar mass are curved. In such a universe, it is difficult to separate the notions of space and matter, because the latter, accumulated in the form of stars or planets, could be regarded as unusual points in space …

Let us consider the observed movements of the planets and try to understand them in Einsteinian space, reduced to three dimensions. To do this, let us imagine a sheet of rubber stretched across a rigid frame. In the centre of the sheet, a billiard ball creates a depression in the surface. We see immediately that the curvature of the surface is very considerable near the ball, and diminishes towards the outside. Let us release a small ball, such as a marble, on the rubber sheet. It will run round the billiard ball, just as the Earth orbits the Sun. In the absence of weight and friction, if we were to release two balls at different distances, they would follow Kepler's Third Law, with the one nearer to the centre moving faster to avoid falling onto the billiard ball.

This is our image, albeit simplified, of the Solar System as seen in the light of relativity. It should be noted that Einstein's theory has enabled us to explain certain local anomalies observed in the Solar System, and which could not be accounted for by Newton's law. In addition, even the curvature of space may be observed indirectly, as we shall see later.

THE ROTATION OF THE PLANETS

The most obvious motion of a planet for an observer located on it is its rotation on its axis, such as shown, on Earth, by the alternation of day and night, and the rapid motion of bodies across the sky. All the

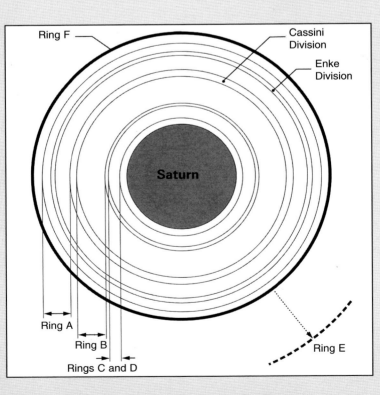

■ THE RINGS OF SATURN, AS DEFINED BY ASTRONOMERS FROM OBSERVATIONS MADE FROM EARTH AND FROM SPACE. RINGS A, B, AND C HAVE BEEN KNOWN FOR MORE THAN A CENTURY, RING E WAS DISCOVERED IN 1980, WITH RING D AND RING F DISCOVERED BY THE PIONEER PROBES IN 1973. THE VOYAGER PROBES RESOLVED THE RINGS INTO THOUSANDS OF CONCENTRIC RINGLETS.

LUNAR ECLIPSES

Date	Magnitude (in %)	Start of partial phase	Start of totality	Mid-eclipse	End of totality	End of partial phase	Notes and zones of visibility
16 May 2003	113	2 h 06	3 h 17	3 h 43	4 h 09	5 h 20	Start visible in Europe, total in South America
09 Nov. 2003	102	23 h 35	1 h 09	1 h 20	1 h 31	3 h 05	Visible in Europe
04 May 2004	130	18 h 49	19 h 53	20 h 30	21 h 07	22 h 11	End visible in W. Europe, total in central Europe
28 Oct. 2004	131	1 h 15	2 h 24	3 h 04	3 h 44	4 h 53	Visible in Europe and America
17 Oct. 2005	6	11 h 34	—	12 h 02	—	12 h 30	Visible solely in the Pacific
07 Sep. 2006	18	18 h 10	—	18 h 54	—	19 h 38	End visible in W. Europe, observable in central Europe
03 Mar. 2007	123	21 h 33	22 h 47	23 h 23	23 h 59	1 h 13	Visible in Europe
28 Aug. 2007	147	8 h 51	9 h 51	10 h 36	11 h 21	12 h 21	Visible solely in the Pacific
21 Feb. 2008	111	1 h 46	3 h 03	3 h 28	3 h 53	5 h 10	Visible in Europe and America
16 Aug. 2008	81	19 h 36	—	21 h 09	—	22 h 42	Start visible in Europe, total in Africa
31 Dec. 2009	7	18 h 57	—	19 h 25	19 h 53	—	Visible in Europe, Africa and Asia
26 Jun. 2010	53	10 h 18	—	11 h 39	—	13 h 00	Visible solely in the Pacific
21 Dec. 2010	125	6 h 34	7 h 41	8 h 17	8 h 53	10 h 00	Start visible in Europe, total in North America
15 Jun. 2011	170	18 h 22	19 h 21	20 h 12	21 h 03	22 h 02	Total, Indian Ocean, India, East Africa
10 Dec. 2011	111	12 h 45	14 h 05	14 h 31	14 h 57	16 h 18	Total, Pacific Ocean, Australia, Japan
15 Apr. 2014	129	5 h 57	7 h 06	7 h 45	8 h 24	9 h 33	Total, Pacific Ocean, USA, Chile, Mexico
08 Oct. 2014	117	9 h 14	10 h 24	10 h 54	11 h 24	12 h 34	Total, Pacific Ocean, Hawaii

TOTAL AND ANNULAR ECLIPSES OF THE SUN

Date	Type	Zone of visibility	Visibility in Europe	Time of maximum	Duration of totality
10 Jun. 2002	Annular	Pacific Ocean		23 h 45	l min 13 s
04 Dec. 2002	Total	Australia, southern Africa		7 h 33	2 min 04 s
31 May 2003	Annular	South Pacific and central America		4 h 10	3 min 37 s
23 Nov. 2003	Total	Antarctica		22 h 47	1 min 57 s
08 Apr. 2005	Annular	South Pacific and central America		20 h 42	42 s
03 Oct. 2005	Annular	Spain, Algeria, Libya, Ethiopia	Partial in Europe	10 h 28	4 min 32 s
29 Mar. 2006	Total	Africa, Middle East, C.I.S.	Partial in Europe	10 h 15	4 min 07s
22 Sep. 2006	Annular	Brazil, South Atlantic		11 h 40	7 min 09 s
07 Feb. 2008	Annular	Antarctica		3 h 55	2 min l4 s
01 Aug. 2008	Total	N. Canada, Greenland, C.I.S.	Partial in Europe	10 h 25	2 min 28 s
26 Jan. 2009	Annular	Indian Ocean, Sumatra, Borneo		7 h 57	7 min 56 s
22 July 2009	Total	India, Nepal, southern Asia		2 h 36	6 min 40 s
15 Jan. 2010	Annular	Central Africa, India, southern Asia		7 h 10	11 min 10 s
11 July 2010	Total	South Pacific		19 h 30	5 min 20 s
04 Jan. 2011	Partial	Europe and north Africa	Partial in Europe	8 h 54	—
20 May 2012	Annular	S. Japan, Pacific, USA		23 h 52	5 min 46 s
13 Nov. 2012	Total	N. Australia, Pacific		22 h 11	4 min 02 s

planets rotate, as do their satellites, although the rotation periods of the latter are often equal to their orbital periods around their parent planets. In most cases, the rotation periods of planets fall into two classes, approximating the terrestrial day or year. We have Earth, 24 hours; Mars, 24 hours; Neptune, 16 hours; Jupiter, Saturn, and Uranus, 10 hours; and then Mercury, 59 days; and Venus, 243 days. The inclination of the axis of rotation goes from 3° for Jupiter to 177° for Venus, via 98° for Uranus. The values for Earth, Mars, Saturn, and Neptune lie around 25°. This inclination is responsible, on Earth and Mars, for significant seasonal effects.

All the planets, except Venus, rotate in the same direction: from west to east. This is why, on Earth, the Sun, the Moon, and all other celestial bodies appear to cross the sky from east to west. Looking in the direction of the Earth's rotation axis (as projected on the celestial sphere), the constellations appear to rotate around a point on the sky, which, for the northern hemisphere, lies in the constellation of Ursa Minor. One star even appears to remain fixed throughout the night and throughout the year. This is the Pole Star, Polaris, which is almost precisely in line with the Earth's axis of rotation. Sailors and hikers are familiar with this star, which indicates north and, by extension, the other points of the compass.

THE ZODIAC, THE ECLIPTIC, AND THE SEASONS

Because all the planets orbit the Sun in almost the same plane, they are always found in the same region of the sky, in the form of a narrow band known as the zodiac. When one studies the positions and the motions of the planets, reference is often made to the ecliptic. This is the great circle that forms the apparent path of the Sun across the sky during the course of a year. It may also be expressed as the projection on the sky of the Earth's orbit around the Sun. This line of the ecliptic appears to oscillate, over the course of a year, from one side of the celestial equator to the other. This is caused by the inclination of the Earth's axis of rotation to the ecliptic. The Sun, in travelling along the ecliptic, reaches four specific points during the year. About 21 March, the Sun passes from the southern hemisphere into the northern. This point, at which the ecliptic crosses the celestial equator, is the vernal (spring) equinox. When the Sun reaches this point (neglecting refraction) it rises exactly in the east and sets precisely in the west. The vernal equinox is the origin of the system of celestial co-ordinates (Right Ascension and Declination), which is, in effect the projection onto the sky of our terrestrial system of longitude and latitude. About 21 June, the Sun, in the northern hemisphere, culminates at $+23°26'$: it is the summer solstice. About 23 September, the Sun crosses the ecliptic from north to south: this is the autumnal equinox. Finally, about 21 December, the Sun reaches the lowest point of the ecliptic, at $-23°26'$: this is the winter solstice. Throughout the course of a year, at any point on the Earth, the Sun therefore varies by $2 \times 23°26' = 46°52'$, which demonstrates the importance of the Earth's axial inclination as regards atmospheric conditions, illustrated by the succession of freezes, droughts, long winter nights, and brilliant summer days that we experience.

It should be noted that the seasons are not the prerogative of Earth; more or less everywhere in the universe, planets must experience winters and springs, although undoubtedly very different from ours. Right next to us there is a striking example: Mars. Like the Earth, it rotates in 24 hours – actually 24^h37^m – and its axis of rotation has a similar inclination (25°11'). Only its year, by virtue of Kepler's Third Law, is rather different: 1 year 322 days. We have found significant seasonal effects on Mars. The temperature at the surface and atmospheric conditions change over the course of a year and, in winter, a thin deposit of frost forms, in places, on its great red deserts.

Over and above these major, but apparently simple and regular, cycles, the planets also experience smaller movements, which are minute on a yearly scale, but which may assume considerable importance over a long course of time. This is the reason for our allusion, earlier in this chapter, to a complex clockwork. For example, although the Earth's rotation axis remains stable as regards its inclination, it describes an extremely slow conical motion relative to the plane of the ecliptic. This motion, known as precession, is similar to that clearly seen with a top, or a gyroscope, whose axis of rotation is inclined to the plane on which it rests. This precession, with a period of about 25 800 years – far different from that of a top – causes the vernal equinox to shift slowly along the ecliptic: this is the famous precession of the equinoxes. Because the vernal equinox is moving along the ecliptic, both the equinoxes and the solstices change with respect to the celestial sphere. The summer solstice, for example, which at present occurs in the zodiacal constellation of Gemini, was in Cancer two thousand years ago.

In addition, the tropics of Cancer and Capricorn, where the Sun culminates at the zenith at the solstices, should really nowadays be called the tropics of Gemini and Sagittarius. These subtle motions, such as precession and nutation, which is a small oscillation of the Earth's axis with a period of 18 years, are caused by gravitational perturbations by the Sun, Moon and planets, which act on the Earth's equatorial bulge. The latter does not, in fact, behave like a homogeneous sphere, because, like the other planets, it is slightly flattened at the poles.

ECLIPSES OF THE SUN

Eclipses are the most spectacular result of the movement of planets and satellites. They occur throughout the universe, between planets, between

■ ONE OF THE MOST IMPRESSIVE ECLIPSES OF THE CENTURY TOOK PLACE ON 11 JULY 1991. IT WAS INCREDIBLY LUCKY FOR ASTRONOMERS: THE PATH CROSSED THE MAUNA KEA OBSERVATORY ON HAWAII, THE GREATEST PRESENT-DAY CENTRE FOR ASTRONOMY. HERE, A SHORT EXPOSURE (1/1000 SEC) REVEALS THE SOLAR CHROMOSPHERE AND PROMINENCES.

satellites, between planets and satellites, and even between stars. We shall take as examples those that we find most striking: eclipses of the Sun and eclipses of the Moon.

As everyone knows, eclipses of the Sun occur when the Earth, the Moon and the Sun are aligned, i.e., when the Moon, as seen from Earth, passes in front of the Sun, hiding the latter for several minutes. But these eclipses reveal an extraordinary factor: the apparent diameters of the Sun and the Moon are practically identical. This arises, of course, because the ratio of their distances, and the ratio of their actual diameters is the same: 1:400. In fact, the diameter of the Moon is just 3476 km, whereas that of the Sun is 1.4 million kilometres.

Thanks to this lucky coincidence, we, living on Earth, have the privilege of observing one of Nature's most spectacular events. Anyone who has seen an eclipse knows how it catches one's imagination, reminding us of the terror felt by people in antiquity, who imagined that they were seeing the Moon engulfing the Sun …

In a simplified version of the Solar System, with planetary orbits that had neither eccentricity nor inclination, there would be a total eclipse of the Sun at each New Moon, practically one per month. In fact, there is roughly one sixth of that number. This is explained by the inclination of the lunar orbit relative to the ecliptic, which causes the former to appear to oscillate from one side of the latter to the other. At each New Moon, our satellite may, therefore, pass above the Sun, below it, or cause a partial eclipse. The apparent equality of the diameters is, in fact, only a first approximation. Because of the eccentricity of both the lunar and terrestrial orbits, it may happen that when a total eclipse is expected, the disk of the Moon does not completely hide that of the Sun. In extreme cases, the Sun's apparent diameter slightly exceeds that of the Moon, and a narrow ring of light remains throughout the total phase. The eclipse is then said to be annular.

The duration of the total phase of a solar eclipse naturally depends on several factors, such as the ratio of apparent diameters and the Moon's actual motion. Let us simply say that it amounts to a few minutes, the record being that of the eclipse of 30 June 1973, which lasted 7 minutes 8 seconds, and that of 11 July 1991, which lasted 6 minutes 58 seconds.

The area involved during a total eclipse, that is, the region swept by the shadow cone of the Moon, is extremely limited, being a few thousand kilometres long, but only a few tens of kilometres wide! In fact, it is extremely rare to observe two total eclipses of the Sun from the same point. At Paris, for example, the interval between the last total eclipse and the next is three centuries. One can understand why scientists and eclipse addicts, often people with a lot of money, undertake long trips to reach a place where the Moon's shadow will pass — even in the middle of the sea! For scientists, an eclipse of the Sun is a unique natural laboratory; it enables them to observe the outer regions of the Sun, which are faint and normally overwhelmed by the light from the true disk of the Sun; to examine the interplanetary medium; or else perhaps to verify the Einstein effect: the deviation of rays of light because of the curvature of space. During a solar eclipse, stars are clearly visible, it is almost like night. At the beginning of the 20th century, Einstein predicted that if a star, apparently extremely close to the eclipsed Sun were observed, the effect of the curvature of space in the neighbourhood of the latter should be detectable, in the form of a slight shift in the position of the star relative to its normal position, caused by a change in the path of the rays of light. Indeed, the Einstein effect has been recorded at numerous eclipses. To the advocates of general relativity, the observation of this effect was a startling demonstration of its validity.

ECLIPSES OF THE MOON

These also occur when the Sun, Earth and Moon are aligned, but in this case, it is the Earth that passes between the Sun and the Moon. Seen from the Moon, therefore, they are eclipses of the Sun!

But to return to Earth. Eclipses of the Sun occur on the day of New Moon. Eclipses of the Moon, occur 180° away, on the day of Full Moon. Seen from the Moon, the apparent diameter of the Earth is more than three times the apparent diameter of the Sun. Because of this, when there is a total eclipse, the whole of the Moon passes inside the Earth's shadow cone. The total phase may last two hours. In addition, the area of the Earth where the eclipse is visible is far larger than in the case of a solar eclipse. As we have seen, the latter is an extremely narrow band, 260 km wide at maximum, whereas lunar eclipses may be seen from anywhere that the Moon is above the horizon during the event, in other words, from a whole hemisphere.

Intuitively, one might imagine that when the Moon passes into the Earth's shadow cone, it would become invisible. But no! In every case, it remains visible. Most often, it becomes a splendid red colour, more or less pronounced, more or less dark. This residual illumination is caused by refraction of the rays of sunlight in the Earth's atmosphere, which then reach the Moon. The variable colour is caused by selective transmission through the atmosphere, red light being absorbed less than the other colours.

Eclipses of the Moon are easily visible with the naked eye, when they offer a marvellous, and always different, spectacle. In addition, this complete change from the normal course of events does not require an expensive expedition, because it is visible from half of the planet. In fact, although the interval between two solar eclipses at a given place may be as much as three centuries, the interval between two total lunar eclipses, is about three years!

TRANSISTS, CONJUNCTIONS AND OCCULTATIONS

There are other phenomena that are comparable to eclipses, at least in principle, if not in scale. Because all the planets move in almost the same plane around the Sun, astronomers are sometimes able to observe the inferior planets, Mercury and Venus, passing in front of the Sun's disk. Because their apparent diameters are respectively, two hundred and thirty times smaller than that of the Sun, they do not cause any obvious change in its brightness. It is also possible to see the amazing spectacle of the

Moon passing in front of a planet. This is known as an occultation. In certain cases, if the occultation is a grazing one – relative to the upper or lower edge of the Moon – through a telescope it is possible to see the planet disappear behind a lunar mountain that appears in profile at the limb, then reappear in a valley, etc., producing an unforgettable spectacle that seems like science fiction. It also frequently happens that the planets appear to cluster together in a small region of the zodiac, giving rise to a planetary conjunction. On the other hand, it is exceptionally rare for one planet to occult another. In more than four centuries of telescopic observation, there has been just one single observation of this phenomenon!

Naturally, nowadays, all these astronomical phenomena are calculated on computers. It is possible to predict an eclipse or occultation to within a few seconds, several centuries in advance.

Another type of occultation that is particularly interesting is the occultation of stars. These naturally allow one to determine with extreme accuracy the motions of the planets, and the diameter of planets or satellites that pass in front of

■ BECAUSE THE EARTH-SUN DISTANCE VARIES, THE APPARENT DIAMETERS OF THE SUN AND MOON MAY VARY SLIGHTLY FROM ONE ECLIPSE TO ANOTHER. HERE, THE DISK OF THE MOON DOES NOT COMPLETELY COVER THAT OF THE SUN. THE ECLIPSE IS SAID TO BE ANNULAR. THE LIGHT FROM THE NON-ECLIPSED PORTION OF THE SUN IS TOO STRONG FOR THE COUNTRYSIDE TO BE PLUNGED INTO DARKNESS.

the star as a function of the length of time during which the light from the latter is extinguished. But they also allow one to discover unsuspected objects that cannot be observed under other conditions. In 1977, Uranus occulted a star. This event revealed two sets of secondary occultations, caused by the presence, hitherto unsuspected, of rings around the planet. Later, in June 1981, the presence of a third satellite of Neptune was established in the same manner.

It is interesting to understand why the motion of the planets in the Solar System was only recognized from the 17th century onward. It is difficult for us, today, to comprehend the differences in the collective understanding between the two periods.

In the days of Copernicus and Kepler, perception of the universe was reduced to that of the celestial sphere, which was a two-dimensional surface, deceptive in appearance, carrying the fixed stars, and against which the planets seemed to move. Only a few scholars had known for a long time that the planets were, in fact, between the stars and us.

Nowadays, a child of six, even before having thought about the question, knows, through images seen on the television, that planets are other worlds, that space is three-dimensional, that the stars are extremely distant, and the planets close by. When, on a clear summer's night, they admire the sky, they know that the appearance of a celestial sphere is an illusion. They no longer ask themselves about the nature of the bodies they admire, but wonder about their chances of going there some day …

T H E P

	km	Semi-major axis (solar radii) a	Distance AU	Eccentricity e	Inclination to the ecliptic i	Longitude of perihelion	Orbital period sidereal P_r
							Orbit
Mercury	57 900 000	83	0.357	0.207	7°.00	76°.67	87.969d
Venus	108 210 000	149	0.723	0.007	3°.39	130°.85	224.701d
Earth	149 600 000	213	1.000	0.017	0°.00	102°.07	365.256d
Mars	227 900 000	328	1.524	0.093	1°.85	335°.97	1 y 321d
Jupiter	778 340 000	1 120	5.203	0.048	1°.31	13°.52	11 y 314d
Saturn	1 427 000 000	2 020	9.5	0.056	2°.49	92°.07	29 y 167d
Uranus	2 869 000 000	4 120	19.218	0.046	0°.77	169°.85	84 y 7d
Neptune	4 490 000 000	6 460	30.11	0.009	1°.78	44°.17	164 y 280d
Pluto	5 966 000 000	8 477	39.80	0.256	17°.14	223°.50	251 y 314d

	Mass Billion billion tonnes M	(Earth = 1)	(Sun = 1)	Density (water = 1) ρ	Kinetic moment $(10^{45} \text{g.cm}^2.\text{s}^{-1})$	Moment of (inertia $MD^2/4$) I	Gravitational acceleration at the equator (cm/s²) g	(Earth = 1) g/gT	Ratio of centrifugal force to g
Mercury	330	0.053	1/6 023 000	5.48	9	—	363	0.38	infinitesimal
Venus	4871	0.816	1/407 700	5.24	185	—	887	0.9	infinitesimal
Earth	5974	1	1/332 946	5.52	267	0.334	978	1	0.0035
Mars	641	0.10	1/3 098 710	3.94	35.2	0.365	373	0.38	0.003 4
Jupiter	1 899 000	317.9	1/1047	1.34	194 000	0.260	2300	2.35	0.084
Saturn	568 000	95.2	1/3498	0.7	78 400	0.210	944	0.96	0.142
Uranus	86 760	14.6	1/22 869	1.30	18 000	0.236	967	0.99	0.1
Neptune	103 000	17.2	1/19 314	1.73	25 000	0.241	1194	1.25	0.022
Pluto	13	0.002	1/152 000 000	2.1	160	—	35	0.04	0.000 0

Abbreviations: IR: infrared. AU: astronomical unit – mean distance between the Earth and the Sun
(1) At quadrature. (2) As seen from the Sun. (3) Maximum. (4) Perihelic opposition. (5) Average temperature at the surface. (6) Average temperature at cloud level.

A N E T S

Orbital period synodic	Velocity (km/s)	Rotation period	Inclination of the equator to the orbit	Equatorial diameter (km)	(Earth = 1)	Flattening optical	Flattening dynamical	Apparent equatorial diameter minimum	maximum
115.88d	47.9	58.646d	2°	4878	0.387	0	0	4"6	12"9
1 y 218d	35.1	243.01d	177°.4	12 101	0.949	0	0	9"8	65"2
—	29.8	23.93h	23°.45	12 756	1	1/347	1/298	—	—
2 y 49d	24.2	24.61h	25°.19	6794	0.533	1/125	1/190	3"6	26"
1 y 33d	13.1	9.83h	3°.1	140 000	10.975	1/16.5	1/15.3	30"	49"
1 y 13d	9.6	10.23h	26°.7	120 600	9.45	1/9.3	1/10.2	15"5	20"5
1 y 4d	6.81	16h	97°.9	50 800	3.9	1/37	1/17	3"7	3"9
1 y 2d	5.4	18.2h	28°.8	48 600	3.81	1/48	1/58	2"2	2"5
1 y 1d	4.7	6.39d	122°	2400	0.188	0	0	0"1	0"2

Critical escape velocity (km/s) V_e	Magnitude m at mean opposition	superior conjunction	Albedo visual % A_v	Bond % A_b	calculated % °C T_c	measured °C T_i	average in IR °C T_m	Atmosphere Principal components	Pressure at surface (millibars)	Magnetic field (order of magnitude)	Known natural satellites (number)
4.3	−0.2	−1.7[1]	6	5	232	340	90[5]	—	0	500	0
10.4	−4.45	−3.6[1]	76	75	−42	462	462[5]	CO_2, N_2	90 000	25	0
11.2	−3.87[2]	—	29	29 to 39	−23	+76[3]	14[5]	N, O_2	1013	45 000	1
5.02	−2.8[4]	+1.8	16	16	−57	+27	−60[5]	CO_2, N_2	2 to 10	5?	2
59	−2.5	−1.7	70	42	−163	−147	—	H_2, He	—	5 000 000	16
36	+0.85	+1.3	70	45	−193	−160	−180[6]	H, He	—	100 000	30
22	5.7	6.0	93	46	−213	−218	−210[6]	H_2, He	—	290 000	20
24	7.8	7.9	84	50	−222	−158	−220[6]	H_2, He	—	100 000	8
1.0	14.5	15	14.5	—	−230	−230	−230	N_2	0	—	1

THE SATELLITES

Name	Mean distance from centre of the planet		Orbit			Diameter (km)	Mass (billion billion tonnes)	Density (water=1)
	(km)	planetary radii	Eccentricity	Inclination	Sidereal orbital period			
SATELLITE OF THE EARTH								
Moon	384 400	60.2	0.054 9	18.3–28.6	27.322	3476	73.4	3.34
SATELLITES OF MARS								
Phobos	9 400	2.76	0.015	1.1	0.318	21.8	0.000 01	2.0
Deimos	23 500	6.91	0.000 8	0.9–2.7	1.262	11.4	0.000 005	1.9
SATELLITES OF JUPITER								
Metis	128 000	1.78	0.0	0.0	0.295	40	0.000 095	—
Adrastea	128 000	1.88	0.0	0.0	0.295	10	0.000 02	—
Amalthea	181 300	2.53	0.002 8	0.4	0.498	220	0.007	—
Thebe	221 000	3.1	0.0	1.25	0.6726	80	0.000 7	—
Io	421 600	5.91	0.0	0.0	1.769	3640	89.2	3.53
Europa	671 000	9.4	0.000 3	0.0	3.551	3130	48.6	3.03
Ganymede	1 071 000	14.99	0.001 5	0.1	7.155	5280	148.9	1.94
Callisto	1 880 000	26.6	0.007 5	0.4	16.689	4840	106.4	1.79
Leda	11 100 000	156	0.148	26.7	240	10	0.000 005	—
Himalia	11 500 000	161	0.158	27.6	250	170	0.009	—
Elara	11 750 000	165	0.207	24.8	260	80	0.000 7	—
Lysithea	11 860 000	166	0.13	29	263	24	0.000 07	—
Ananke	21 250 000	294	0.17	147	617	20	0.000 03	—
Carme	22 500 000	313	0.21	164	692	30	0.000 09	—
Pasiphae	23 500 000	329	0.38	145	739	40	0.001 9	—
Sinope	23 700 000	332	0.28	153	758	28	0.000 07	—
SATELLITES OF SATURN								
Pan	133 500	2.25	0.0		0.495	10	—	—
Atlas	137 000	2.27	0.0		0.599	15	—	—
Prometheus	139 400	2.31	0.0		0.611	100	—	—
Pandora	141 700	2.34	0.0		0.627	100	—	—
Epimetheus	151 400	2.51	0.01	0	0.692	80	—	—
Janus	151 400	2.51	0.01	0	0.698	80	—	—
Mimas	188 000	3.12	0.020	1.5	0.942	390	0.038	1.17
Enceladus	240 000	3.98	0.004 4	0.02	1.370	500	0.084	1.24
Tethys	295 000	4.9	0.002 2	1.08	1.887	1050	0.64	1.26
Telesto	294 700	4.9	0.0	—	1.887	30	—	—
Calypso	294 700	4.9	0.0	—	1.887	20	—	—
Dione	377 500	6.26	0.002 2	0.02	2.236	1120	1	1.44
Helene	378 000	6.27	0.01	0.2	2.74	30	—	—
Rhea	527 000	8.7	0.000 9	0.35	4.517	1530	2.50	1.33
Titan	1 222 000	20.3	0.029	0.2	15.94	5140	137	1.94
Hyperion	1 484 000	24.7	0.104	0.3	21.276	300	—	—
Iapetus	3 562 000	59.3	0.028	14.72	79.33	1440	1.90	1.20
Phoebe	12 960 000	216	0.163	150.05	550.45	200	—	—

Name	Mean distance from centre of the planet		Orbit			Diameter (km)	Mass (billion billion tonnes)	Density (water=1)
	(km)	planetary radii	Eccentricity	Inclination	Sidereal orbital period			
SATELLITES OF SATURN DISCOVERED IN 2000								
S/2000 S1	23 337 256	387.02	0.375	172.7	1328	—	—	—
S/2000 S2	15 000 000	248.76	0.462	45.9	685	—	—	—
S/2000 S3	18 251 000	302.67	0.380	48.5	917	—	—	—
S/2000 S4	17 952 000	297.71	0.613	34.9	900	—	—	—
S/2000 S5	11 370 000	188.55	0.166	48.4	450	—	—	—
S/2000 S6	11 370 000	188.55	0.359	49.2	452	—	—	—
S/2000 S7	19 897 000	329.97	0.565	174.9	1053	—	—	—
S/2000 S8	15 409 000	255.54	0.212	148.7	719	—	—	—
S/2000 S9	18 401 000	305.16	0.212	169.8	933	—	—	—
S/2000 S10	17 802 000	295.22	0.609	34.5	887	—	—	—
S/2000 S11	16 456 000	272.90	0.456	37.4	788	—	—	—
S/2000 S12	19 149 000	317.56	0.144	174.7	992	—	—	—
SATELLITES OF URANUS								
Cordelia	50 000	1.95	0.000 47	0.19	0.335	25	—	—
Ophelia	54 000	2.11	0.010	0.09	0.376	30	—	—
Bianca	59 000	2.30	0.000 88	0.16	0.435	40	—	—
Cressida	62 000	2.4	0.000 23	0.04	0.464	60	—	—
Desdemona	63 000	2.46	0.000 23	0.16	0.474	55	—	—
Juliet	64 000	2.50	0.000 58	0.06	0.493	80	—	—
Portia	66 000	2.58	0.000 16	0.09	0.513	100	—	—
Rosalind	70 000	2.80	0.000 1	0.28	0.558	50	—	—
Belinda	75 000	2.93	0.000 11	0.03	0.624	65	—	—
Puck	86 000	3.36	0.000 05	0.31	0.761	150	—	—
Miranda	130 000	5.1	0.017	3.4	1.41	480	0.17	1.35
Ariel	192 000	7.6	0.002 8	0.0	2.52	1180	1.6	1.65
Umbriel	267 000	10.5	0.003 5	0.0	4.144	1220	1	1.50
Titania	438 000	17.2	0.002 4	0.0	8.706	1620	5.9	1.68
Oberon	587 000	23.1	0.00	0.12	13.4	1570	6.0	1.60
Caliban	7 230 000	285	0.159	140.89	597.47	80	—	—
Stephano	7 979 000	314	0.228	144.01	673.56	—	—	—
Sycorax	12 178 000	479	0.523	159.40	1283.27	160	—	—
Prospero	16 665 000	656	0.439	151.75	2037.14	—	—	—
Setebos	17 879 000	704	0.551	158.03	2273.34	—	—	—
SATELLITES OF NEPTUNE								
Naiad	48 200	1.94	0.000 38	4.74	0.294	55	—	—
Thalassa	50 000	2.00	0.000 23	0.21	0.311	80	—	—
Despina	52 500	2.12	0.000 17	0.07	0.335	150	—	—
Galatea	61 900	2.50	0.000 07	0.05	0.429	160	—	—
Larissa	73 600	2.97	0.001 38	0.2	0.555	200	—	—
Proteus	117 600	4.74	0.000 46	0.55	1.122	400	—	—
Triton	355 500	14.6	0.000 02	159.9	5.876	2700	20	2.0
Nereid	5 355 000	228	0.74	27.7	359.8	300	—	—
SATELLITE OF PLUTO								
Charon	19 000	15.80	0	94	6.39	1200	1	1.4

TELESCOPES AND SPACEPROBES

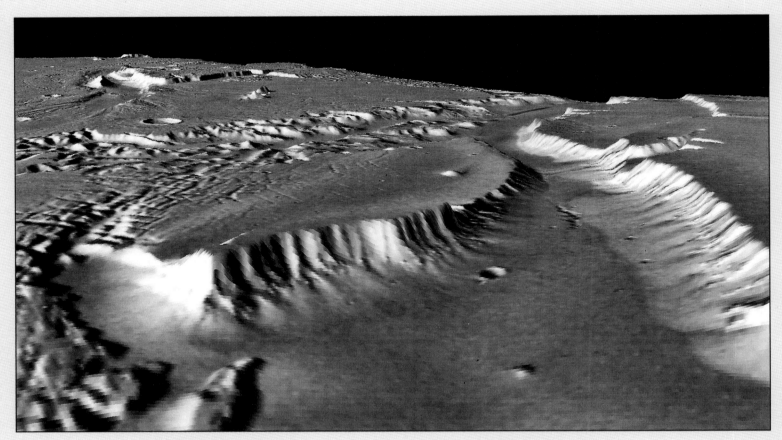

■ THIS LOW-LEVEL IMAGE OF A MARTIAN VALLEY IS PARTIALLY IMAGINARY. THIS RELIEF IMAGE HAS BEEN CREATED USING A COMPUTER, FROM ELECTRONIC IMAGES OBTAINED BY THE VIKING ORBITERS IN 1976. THE PROGRESS IN IMAGE-PROCESSING SOFTWARE NOWADAYS ALLOWS EXPERTS TO WORK ON OLD IMAGERY.

Well before the industrialization of the 19th century multiplied machines, filled the sky with smoke, and brought the use of artificial light into widespread use – all things that weaken and extinguish the feeble twinkling of the stars – people were permanently confronted with the sky and the celestial bodies in it. Those objects that were fixed, appeared to be more or less permanent; those that wandered about, traced their paths against a sky, appeared easy to interpret. At first, astronomy was practised with the naked eye, then with the aid of passive instruments, that allowed one to determine either the time objects passed a specific point in the sky, or else their altitude above the horizon. Very quickly, the Sun, Moon and planets became favoured objects. The bright light from the Sun or the pallid light from the Moon governed peoples' lives, and eclipses, which were sometimes predicted by astrologers or soothsayers, were both superb and terrifying, revealing the divine nature of the world. The planets, with their wondrous motion across the heavens, their variations in brightness, the strange nature of their light, which was fixed, contrasting with the scintillation of the stars, also seemed to possess some divine essence. The personality of each of these wandering bodies reflected that of a god: Mars, with its bloody light, the spirited god of war; Venus, with its pure and brilliant light, modestly veiled by the light of dawn, the goddess of love; and Jupiter, whose motions on the celestial sphere and strong, even light evoked the calm power of the king of the gods.

This view prevailed until the 17th century and a starry night in January 1610, when, for the very first time, Galileo turned a telescope towards the stars. This was the first revolution in observational astronomy, which thus entered the modern age.

THE REFRACTORS' GOLDEN AGE

The advent of the astronomical telescope radically altered our view of those other worlds. The planets ceased to be simply points of light and became other Earths, with complex, changing surfaces, sometimes accompanied by multiple satellites. From then on, the progress of astronomy was closely linked with that of the optical industry. The astronomical refractor, ideally suited for the observation of the planets, held sway for three centuries, from the time of Galileo until the beginning of the 20th century. The object glass, the key piece of this instrument, was slowly perfected; its diameter, which determines the brightness and the delicacy of detail that is detectable, grew from 5 cm in 1650 to 50 cm around 1850. A few giant refractors were constructed in the last decade of the 19th century, and were never surpassed subsequently, for reasons that we will discuss later. These 'monsters' were the Grande Lunette at Meudon, near Paris, whose object glass measures 83 cm in diameter, and the refractors at Lick and Yerkes, in the United States, with apertures of 89 and 101 cm (35 and 40 inches), respectively.

PHOTOGRAPHY

A second revolution occurred in the study of astronomy when the eye of the astronomer was replaced by the photographic plate. Mastery of this technique, and all those that flowed from it, took a very long time, and astronomers worked for decades before beginning to benefit from it. The first photographs of the Moon date from 1840. The planets started to be photographed at the end of the 19th century. In parallel with this, the finest tool of physics, spectroscopy and its ancillary techniques, made its appearance. A spectrograph is literally a means of scanning light. It decomposes light into its primary colours, and allows one to study the precise physical characteristics of the radiating body. Observation of the light from a celestial object, wavelength by wavelength, allows one to calculate the temperature and density of a medium, its chemical composition, the velocity at which its particles are moving, etc. In the case of the planets, photometric, spectrographic, and polarimetric analysis allows one to determine the composition of planetary atmospheres, the rotational velocity, atmospheric

■ ONE OF THE MOST POWERFUL INSTRUMENTS FOR SOLAR OBSERVATION. THIS 50-CM REFRACTOR IS LOCATED AT THE PIC DU MIDI OBSERVATORY. PROTECTED BY A SPECIAL DOME FROM THE SUN'S HEAT, IT IS ALSO COOLED SO THAT THE LIGHT FROM THE SUN DOES NOT INCINERATE THE CINE CAMERA INSTALLED AT ITS FOCUS.

■ FOR NEARLY A CENTURY, THE BEST PHOTOGRAPHS OF THE PLANETS HAVE BEEN OBTAINED AT THE PIC DU MIDI OBSERVATORY, IN THE PYRENEES. AT AN ALTITUDE OF 2870 M, THIS ISOLATED PEAK, LYING ON THE NORTHERN SIDE OF THE MOUNTAIN RANGE, ENJOYS EXCEPTIONALLY CLEAR, CALM ATMOSPHERIC CONDITIONS. THE 1-M TELESCOPE IS RECOGNIZED WORLD-WIDE FOR THE QUALITY OF ITS RESULTS.

motions, and (for the terrestrial planets) the surface temperature, the roughness, and the nature of their surfaces.

MODERN TIMES

Nowadays, the face of astronomical observation has changed. Reflectors, which are more powerful and gather more light, have replaced the 'long refractors that scare people' as Molière put it. Detectors have diversified and, above all, are capable of recording the image of astronomical objects over an extended range of wavelengths: ultraviolet, visible, infrared, and radio. When the Earth's atmosphere prevents this radiation from being detected, telescopes are mounted on balloons, on aircraft, on research rockets, or on satellites, as was

done with the Hubble Space Telescope. To give a few examples: the rings of Uranus were discovered in 1977 from an aircraft, the Kuiper Airborne Observatory. New satellites and rings of Saturn were discovered in 1980, using ultrasensitive photographic films, electronic cameras, and also new instruments, charge-coupled devices (CCDs), which are photosensitive electronic matrices. Each of the sensitive elements of a CCD chip records the light reaching it, directing it to a computer which reconstitutes the overall image and stores it in digital form. A CCD matrix consists of about one million of these 'pixels' (picture elements). Since the beginning of the 1980s, most reflectors have been fitted with CCD cameras.

There is one point in common for all planetary observations: they are primarily made from high-altitude observatories. The atmospheric purity, transparency, and stability necessary for planetary observation are found only at altitude, where the air is rarefied, and low in water vapour and aerosols. A few observatories across the world have become noted for the quality of seeing that they offer astronomers. The most famous are the Pic du Midi, in the Pyrenees; Catalina, in Arizona; and Lick, in California. We may also include Mauna Kea, at an altitude of 4200 m on Hawaii, where there are already four telescopes of between 3 and 10 m in diameter. Finally, the region of the electromagnetic spectrum that is observed from the ground has been extended to radio wavelengths, thanks to radio astronomy and radar-astronomy, the latter consisting of detecting the echo of a powerful radio pulse emitted by a radio telescope that is

PRINCIPAL INTERP

Mission	Nation	Launch	End of the mission	Principal goals
Luna 1	USSR	1959	1959	Lunar fly-by
Luna 2	USSR	1959	1959	Lunar impact
Luna 3	USSR	1959	1959	Photos of lunar farside
Mariner 2	USA	1962	1962	Venus fly-by
Ranger 7	USA	1964	1964	Lunar photographs
Mariner 4	USA	1964	1967	Mars: fly-by and photos
Ranger 8	USA	1965	1965	Lunar photographs
Ranger 9	USA	1965	1965	Lunar photographs
Zond 3	USSR	1965	1965	Photos of lunar farside
Luna 9	USSR	1966	1966	Lunar landing and photos
Surveyor 1	USA	1966	1967	Lunar landing and photos
Lunar-Orbiter 2	USA	1966	1967	Lunar photographs
Venera 4	USSR	1967	1967	Venus: atmospheric capsule
Venera 5	USSR	1969	1969	Venus: atmospheric capsule
Venera 6	USSR	1969	1969	Venus: atmospheric capsule
Mariner 6	USA	1969	1969	Mars: fly-by and photos
Mariner 7	USA	1969	1969	Mars: fly-by and photos
Venera 7	USSR	1970	1970	Landing on Venus
Luna 16	USSR	1970	1970	Lunar sample return
Luna 17	USSR	1970	1970	Lunar robot Lunokhod 1
Mars 2	USSR	1971	1972	Landing on Mars
Mariner 9	USA	1971	1972	Mars: fly-by and photos
Luna 19	USSR	1971	1971	Lunar landing
Luna 20	USSR	1972	1972	Lunar sample return
Pioneer 10	USA	1972	on-going	Jupiter fly-by
Venera 8	USSR	1972	1972	Landing on Venus
Luna 21	USSR	1973	1973	Lunar robot Lunokhod 2
Pioneer 11	USA	1973	on-going	Jupiter and Saturn: fly-bys
Mars 5	USSR	1973	1974	Photos of Mars
Mariner 10	USA	1973	1975	Mercury, Venus: fly-bys
Luna 22	USSR	1974	1975	Lunar landing

ANETARY MISSIONS

Mission	Nation	Launch	End of the mission	Principal goals
Venera 9	USSR	1975	1975	Venus: landing and photos
Venera 10	USSR	1975	1975	Venus: landing and photos
Viking 1	USA	1975	1982	Mars: orbit and landing
Viking 2	USA	1975	1978	Mars: orbit and landing
Voyager 1	USA	1977	on-going	Fly-bys of Jupiter and Saturn
Voyager 2	USA	1977	on-going	Jupiter, Saturn, Uranus, Neptune
Pioneer Venus 1	USA	1978	1992	Venus: radar mapping
Pionner Venus 2	USA	1978	1978	Venus: atmospheric capsule
Venera 11	USSR	1978	1978	Venus: landing
Venera 12	USSR	1978	1978	Venus: landing
Venera 13	USSR	1981	1982	Venus: landing and photos
Venera 14	USSR	1981	1982	Venus: landing and photos
Venera 15	USSR	1983	1984	Venus: radar mapping
Venera 16	USSR	1983	1984	Venus: radar mapping
Vega 1	USSR	1984	1985	Venus and Comet Halley
Vega 2	USSR	1984	1985	Venus and Comet Halley
Sakigake	Japan	1985	1986	Fly-by of Comet Halley
Giotto	Europe	1985	1992	Photos of Comet Halley
Magellan	USA	1989	1994	Venus: radar mapping
Galileo	USA	1989	2002	Fly-bys of Gaspra, Ida, Jupiter
Ulysses	Europe	1990	on-going	Solar polar orbit
Clementine	USA	1994	1994	Lunar photography
Mars Pathfinder	USA	1996	1997	Mars: landing
Near	USA	1996	2001	Asteroid fly-by and landing
Cassini–Huygens	Europe–USA	1997	in progress	Saturn: fly-by and landing on Titan
Mars Global Surveyor	USA	1997	in progress	Mars: mapping
Mars Odyssey 2001	USA	2001	in progress	Mars: mapping

This list of the principal planetary mission is not exhaustive. In particular, it excludes all the Apollo manned lunar missions, and a certain number of missions that failed, or did not contribute significantly to knowledge of the Solar System. Similarly, only the most important aspect, within the context of this book, has been mentioned: namely imagery of the bodies concerned. It should be noted, however, that each spaceprobe actually recorded thousands of observations other than photographs, such as measurements of magnetic fields, chemical compositions, etc.

returned by a planet. This very promising technique has seen considerable success since the beginning of the 1990s. In this connection we may note the probable discovery of ice at both of Mercury's poles; imagery of the asteroids Castalia and Toutatis; and research into the state of Titan's surface.

TRIPS TO THE PLANETS

Naturally, planetary astronomy has greatly benefited, in the last thirty years, from the extraordinary contribution made by space missions. All the planets in the Solar System, except Pluto, have been visited by human-built robots. Spaceprobes are equipped with small telescopes, whose power is comparable with either a powerful telephoto lens, or a small amateur telescope. During their journey to their objective, which very often lasts several years, they orientate themselves in space using stars as reference points. Stars, because of their enormous distance, are considered as fixed and unchanging references. From time to time, the spaceprobes' trajectories are corrected slightly by very short bursts from small, gas thrusters. Spaceprobes are in constant contact with the Earth. The engineers communicate with the spacecraft through giant radio aerials – most of which are American. Those used

by NASA's Jet Propulsion Laboratory are between 34 and 70 m in diameter, and are located in California, Spain, and Australia, so that, despite the Earth's rotation, the spaceprobe is always accessible from one of the receiving stations. The large size of these stations and the power of their antennae is warranted because of the distance of the spaceprobes – sometimes several billion kilometres – and the weak transmitters that the spaceprobes carry. The transmitting antennae on board the spacecraft measure just a few decimetres to about 3 m across. In fact, in certain critical cases, such as for Voyager 2's encounter with Neptune in 1989, NASA's antennae were supplemented by even more powerful networks of telescopes, such as the VLA (Very Large Array) in New Mexico. Similarly, in 1986, the European Giotto spaceprobe, which had a high-speed encounter with the nucleus of Comet Halley, also benefited from international help in navigation.

Like the telescopes at terrestrial observatories, those on spaceprobes are equipped with CCD cameras, which need to be turned automatically towards the object being studied. Aiming a telescope in real time is, in practice, impossible, because communications between Earth and probe are, despite their high data-rate, carried at the speed of light and are subject to delays ranging from a few minutes to several hours. An order given on Earth, for example, at a particular instant, might not reach the spaceprobe until an hour later. In general, therefore, the engineers program the on-board computer several days in advance of the encounter between the probe and a planetary system, so that the telescopes and various scientific instruments follow a pre-programmed sequence of observations.

DATA-PROCESSING

In less than thirty years, from the very early steps in the conquest of the Moon to the triumph of the Magellan probe in

scanning Venus, spaceprobes, mostly American, but also Soviet and European, have revealed sixty-odd new worlds: some twenty-odd of which are true planetary bodies, with volcanoes, glacial valleys, peaks 10 000 m high, plateaux saturated with craters, and breathtaking canyons. The harvest has been several million electronic images, which have had to be examined, classified, analyzed, and finally archived. Once the first moments of discovery have passed, researchers had to start mapping these new planets, some of which were hardly much larger than that described in Saint-Exupéry's famous children's book, *The Little Prince*. Cartography, covering everything from surveying a field of maize to determining the height of all the peaks in the Himalaya, is one activity to which every administration has devoted a great deal of investment, with greater or lesser success, ever since the Earth was regarded as flat. The United States' government entrusted the responsibility of mapping all the planets in the Solar System to a very small unit of the US Geological Survey, at Flagstaff in Arizona. Or, to be more precise, all the planets that could be mapped, which excludes the planets that have no geography: Jupiter, Saturn, Uranus, and Neptune. We are left with some thirty-odd planets

■ THE HUBBLE SPACE TELESCOPE, JUST BEFORE LAUNCH, IN 1990. SINCE THEN, THE HUBBLE TELESCOPE HAS BEEN VISITED THREE TIMES BY AMERICAN ASTRONAUTS.

■ THE GALILEO PROBE, BEFORE ITS DEPARTURE FOR JUPITER, IN 1989. IT WENT INTO ORBIT AROUND THE GIANT PLANET IN 1995, AND WAS STILL OPERATING IN EARLY 2002.

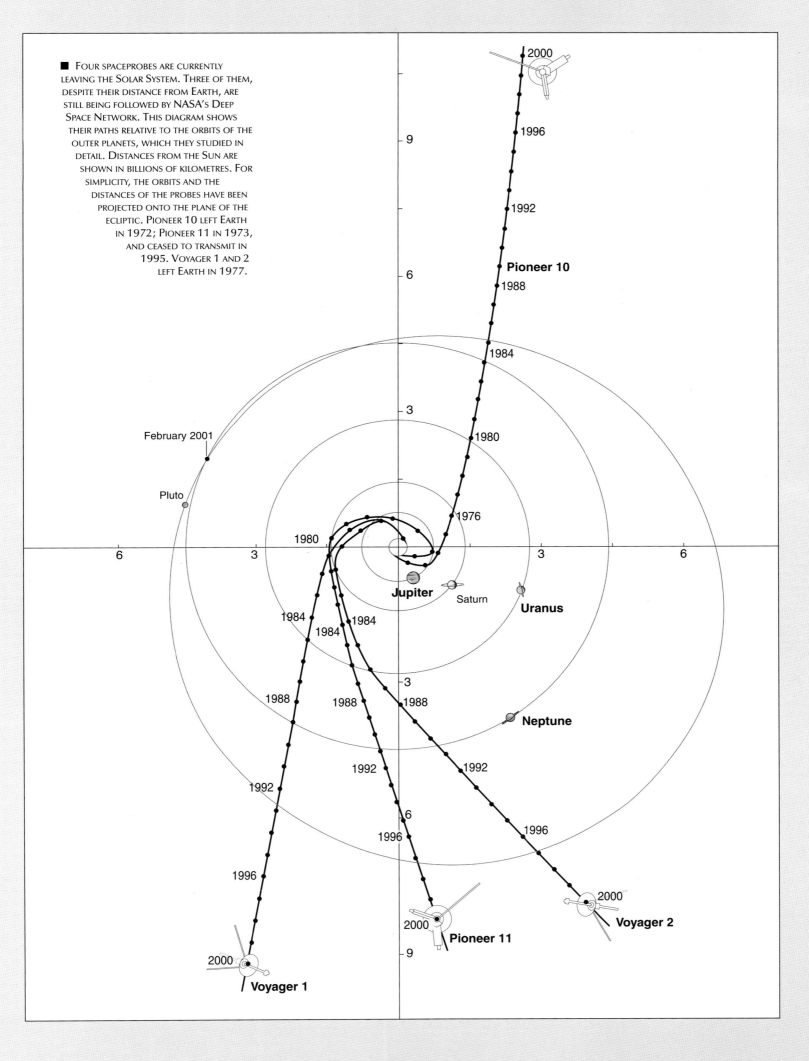

■ FOUR SPACEPROBES ARE CURRENTLY
LEAVING THE SOLAR SYSTEM. THREE OF THEM,
DESPITE THEIR DISTANCE FROM EARTH, ARE
STILL BEING FOLLOWED BY NASA'S DEEP
SPACE NETWORK. THIS DIAGRAM SHOWS
THEIR PATHS RELATIVE TO THE ORBITS OF THE
OUTER PLANETS, WHICH THEY STUDIED IN
DETAIL. DISTANCES FROM THE SUN ARE
SHOWN IN BILLIONS OF KILOMETRES. FOR
SIMPLICITY, THE ORBITS AND THE
DISTANCES OF THE PROBES HAVE BEEN
PROJECTED ONTO THE PLANE OF THE
ECLIPTIC. PIONEER 10 LEFT EARTH
IN 1972; PIONEER 11 IN 1973,
AND CEASED TO TRANSMIT IN
1995. VOYAGER 1 AND 2
LEFT EARTH IN 1977.

2000

1996

1992

Pioneer 10
1988

1984

1980

1976

February 2001

Pluto

1980

Jupiter Saturn **Uranus**

1984 1984 1984

3

1988 1988 1988

Neptune

1992 1992

1992

1996 1996

1996

2000

2000 **Voyager 2**

Pioneer 11

1996

2000 **Voyager 1**

and satellites whose diameters exceed 200 km, well over thirty asteroids with diameters of between 200 and 1000 km, and thousands of smaller objects, small satellites, comets, and asteroids. Decades of secure employment for the small team at the USGS Planetary Data Facility.

Mapping the planets: for planetologists, such a task is absolutely crucial. First, because an accurate map represents the basic reference for everything. Second, because a simple map, is, in itself, a scientific tool. One researcher will use it for years to study the cratering of one region relative to another, which will enable them to reconstruct the history of the planet. Another, by studying the number of volcanoes, their diameters, and their altitudes, will be able to work out the bodies' internal mechanisms. Finally, the maps that the USGS currently creates are often – depending on how well the body being studied is known – specialized. There are albedo maps, geological maps, or topographical maps. They enable planetologists the world over to determine, at little cost, the regions that are of particular interest for their own studies, and then, if necessary, enables them to ask NASA for the original material.

The specialists at the USGS have chosen an original method of depicting the mapping of bodies in the Solar System: shaded relief. This drawing technique, consists of representing the planetary surface with a slight, imaginary shadow, as if the whole surface were lit by sunlight

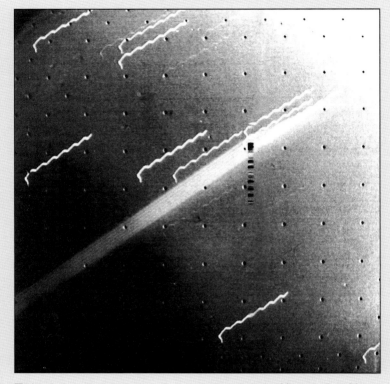

■ A RAW IMAGE AS TRANSMITTED BY THE PLANETARY PROBE VOYAGER 1 IN MARCH 1979. THE PHOTOGRAPH SHOWS, IN THE CENTRE, JUPITER'S NARROW RINGS. THE ZIGZAG LINES ARE STARS, WHICH 'MOVED' DURING THE EXPOSURE WHICH LASTED MORE THAN 11 MINUTES. THE GRID OF DOTS IS USED FOR CALIBRATION PURPOSES.

from the west. Then, region by region, and dependent on the scale chosen – 1/1 000 000 for the Moon and Mars, for example – the cartographers form an overall compilation of the video images

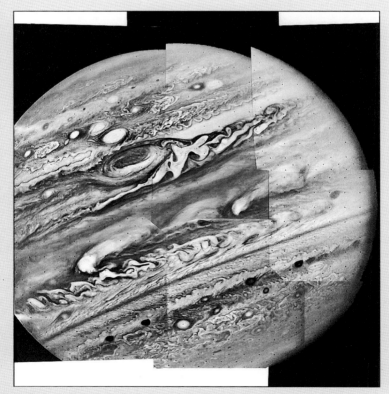

■ A RAW IMAGE OF THE PLANET JUPITER, OBTAINED BY THE VOYAGER 1 PROBE IN FEBRUARY 1979. THIS PHOTOGRAPH IS A MOSAIC CREATED FROM 9 ELECTRONIC IMAGES. THE CALIBRATION GRID MAY BE REMOVED IN THE IMAGE-PROCESSING STAGE. SIMILARLY IT IS POSSIBLE TO ARTIFICIALLY ADJUST THE BRIGHTNESS OF EACH IMAGE.

transmitted back by the probe or probes. The work, which is long and complex, consists of correcting each image to take into account, first, the viewing angle, then the orbital position of the spaceprobe, and finally, the chosen projection: Mercator projection for the equatorial regions, and stereographic for the poles. The draughtsmen, before they start the actual cartography, need to interpret the images that they have in front of them. Some are full of detail, but are rendered almost illegible because of transmission errors or background noise. In others the relief is hidden by long shadows cast by the early-morning light, or else rendered undetectable by harsh, overhead lighting. Still other images may be affected by major distortions of perspective.

For this work as a whole, the eye of the draughtsman, when well trained, is far more efficient (and far quicker) than even the most sophisticated computer. All the planetary draughtsmen at the USGS use air-brushes. They start by depicting the most obvious features – such as mountains and craters – on tracing paper laid over an original photograph that has been rectified by the computer. Then they add the fine details, balancing the shadows and, finally, for certain charts, add the albedo markings.

To carry out a proper test of the validity of this method, the researchers at the USGS used specific photographs of Mars obtained by the Mariner 9 probe. Airbrushed maps of Mars were prepared in this way, and then compared with the far more detailed images

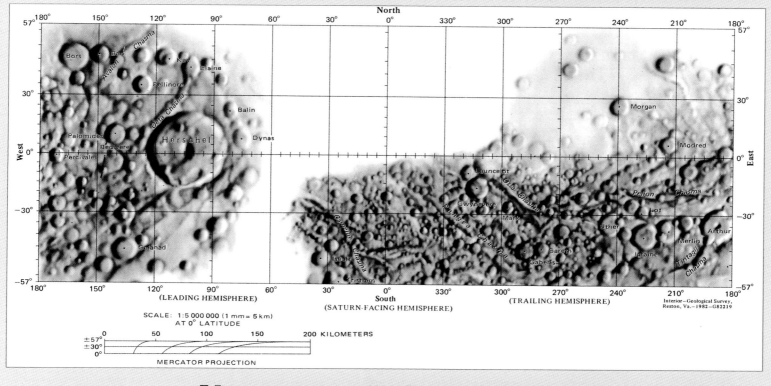

North

West — East

South

SCALE: 1:5 000 000 (1 mm = 5 km)
AT 0° LATITUDE

(LEADING HEMISPHERE)
(SATURN-FACING HEMISPHERE)
(TRAILING HEMISPHERE)

Interior—Geological Survey,
Reston, Va.—1982—G82219

MERCATOR PROJECTION

■ THE OFFICIAL MAP OF THE SATELLITE MIMAS. THIS SMALL SATELLITE OF SATURN WAS NOT COMPLETELY COVERED BY THE CAMERAS CARRIED BY THE TWO VOYAGER PROBES, WHICH ACCOUNTS FOR THE WHITE AREAS ON THE MAP. THIS WAS CREATED WITH AN AIRBRUSH, BY THE EXPERTS AT THE UNITED STATES GEOLOGICAL SURVEY, AT FLAGSTAFF IN ARIZONA.

obtained by the two Viking probes. The agreement between the maps and the images was complete: in interpreting the Mariner images, the cartographers had neither deformed the details, nor, more importantly, added information! But this still does not prevent some users of the USGS maps from committing errors of interpretation. The inevitable inaccuracies in the data, in particular for the satellites of the giant planets, leads the USGS to publish their maps with the prudent description of 'preliminary'.

The USGS also has to cope both philosophically and with good humour with the unexpected events that occur in space research, which sometimes lead to considerable disappointments. In 1980, for example, excitement was at its height at Flagstaff, when Voyager was racing at 100 000 km/h and without the slightest technical problem, towards the mysterious world of Titan. As it brushed past Saturn's largest satellite, Voyager should have been able to return incredibly clear images of Titan's landscape. Alas, it was not to be! The probe worked to per-

fection, but all that the telescopes revealed to the disappointed cartographers and planetologists was a world completely covered in clouds.

Planetary cartography is not the only activity of the Planetary Data Facility. At Flagstaff, they also directly process the images obtained by NASA's spaceprobes. It is only now, with computers, that it is possible to make the most of the photographs of Mars that were obtained in 1976 by the Viking probes. Using their VDUs, the experts at the USGS manipulate images obtained through multiple colour filters, eliminate background noise, remove optical and electronic artefacts, and finally create true- or false-colour photographs of the most interesting regions of Mars.

A very large fraction of the photographs reproduced in this book are the result of work by the USGS. Their martian training is standing these true scientific artists in good stead as they begin to work on the fantastic harvest from the Magellan probe, which, between 1990

and 1993, covered practically 100 per cent of the surface of Venus. This is, in prospect, a colossal work: for the first time, Venus was photographed (by radar, to be exact) with a resolution of about 100 to 200 m. There are thus thousands of images that must be collated, provided with a proper coordinate grid, rectified, processed, and coloured. There is more: for certain regions of Venus, the experts of the USGS can even work on relief images. What is at stake is a wonderful archive for planetologists, and a beautiful map of Venus unveiled, covered in mountains and craters that one day must be named. After the wave of data from Venus, the cartographers at the USGS will have to turn again to the red planet. Several martian missions are currently in operation, such as Mars Global Surveyor, or in prospect, and new maps of Mars will have to be prepared. In preparation for the far-off day, when the USGS will be able to give a beautiful folding map to the astronauts as they set off for the red planet.

■ THIS LAST-QUARTER MOON WAS PHOTOGRAPHED BY GÉRARD THÉRIN USING A REFLECTOR WITH AN APERTURE OF 225 MM AND A FOCAL LENGTH OF 2.7 M. ALONG THE LUNAR TERMINATOR, THE LINE AT WHICH THE SUN IS SETTING, THE RELIEF IS GREATLY ACCENTUATED BY THE GRAZING ILLUMINATION.

AMATEUR OBSERVATION OF THE PLANETS

Astronomy has undoubtedly been practised since the dawn of time. The Magdalenian artist, on leaving the darkness of the sanctuary at Lascaux after having painted the wonderful scene of the hunter and the aurochs, must have lifted his eyes to a sky full of stars. From the Renaissance period, and Galileo's first experiments, astronomy, which was practised by rich scholars, who had ever more powerful telescopes built for them, slowly spread across Europe. Was Newton – that inspired mathematician and a brilliant jack-of-all-trades when it came to knowledge, as well as being clever with his hands – a 'professional' or an 'amateur' when he first proposed a type of telescope that used a mirror? The history of astronomy is full of amateurs who played a large part in the progress of the science. William Herschel, when he discovered Uranus in 1781, was earning his living by giving music lessons. Clyde Tombaugh discovered Pluto in 1930, after having drawn the changing appearance of Mars throughout his adolescence, which he passed on a farm. Then there was Charles Boyer, a modest amateur observing the sky from Brazzaville, who discovered, long before his professional colleagues, the four-day rotation of the atmosphere of Venus. Observing the sky became one of the fashionable hobbies for a gentleman at the end of the nineteenth century when astronomy – which was popularized, particularly in France, by Camille Flammarion – had acquired the status of a modern science. At that time, planets undoubtedly attracted more attention from the general public than they do nowadays. Not without reason: the planets, and Mars in particular, which was kept under intense scrutiny by zealous and wildly imaginative observers, seemed to be inhabited.

The Martians, invented at Flagstaff by Percival Lowell, an amateur and millionaire, were sacrificed on the altar of science by another amateur, Eugène Antoniadi, at Meudon.

Freedom for imagination and an open mind, constant accessibility, ready availability and, above all, the lack of any obligation to obtain a result have allowed amateur astronomers today to acquire a status that would be the envy of many amateur scientists involved in other disciplines. There are two other basic reasons for this: on the one hand, the practice of astronomy poses no threat to the environment and, being therefore free, it is also completely without repercussions – unlike subjects such as archaeology, geology, or palaeontology. On the other hand, the universe is too vast for professional astronomers alone, and unexplored fields of planetary and stellar investigation are endless and extend across the whole sky. An amateur 'takes nothing' from a professional, and the rays of light that enter their telescopes are utterly independent and do not interfere in the slightest with one another. One of astronomy's other strengths is that whether the universe is observed with a 10-metre telescope or with the naked eye, it still retains the ability to surprise us with its beauty and its richness. Amateur astronomers can choose, like professionals, to specialize, to opt for larger and larger apertures, or the most powerful optics, or, in quite the opposite fashion, to choose the freedom, the flexibility, and the pleasure of a night's observation with no specific purpose.

■ PHOTOGRAPHS OF THE MOON OBTAINED BY THE BEST AMATEUR ASTRONOMERS RIVAL THOSE OF PROFESSIONAL ASTRONOMERS. THIS EXCEPTIONALLY HIGH-QUALITY IMAGE OF THE GREAT BASINS IN THE REGION OF PTOLEMY WAS OBTAINED BY CHRISTIAN ARSIDI, FROM THE PARIS SUBURBS, WITH A REFLECTOR, 300 MM IN DIAMETER AND 15 M FOCAL LENGTH.

OPTICAL CHARACTERISTICS OF SOLAR-SYSTEM SATELLITES

Satellite	Discoverer	Year of discovery	Mean apparent magnitude	Mean apparent diameter
EARTH				
Moon	—	—	−10	1800″
MARS				
Phobos	A. Hall	1877	11.3	0.1″
Deimos	A. Hall	1877	12.4	0.05″
JUPITER				
Metis	S. Synnot	1979	17.5	0.01″
Adrastea	D. Jewitt	1979	18.7	0.001″
Amalthea	E. Barnard	1892	14.1	0.07″
Thebe	S. Synnot	1979	16.0	0.003′
Io	Galileo	1610	5.0	1.20″
Europa	Galileo	1610	5.3	1.05″
Ganymede	Galileo	1610	4.6	1.70″
Callisto	Galileo	1610	5.6	1.60″
Leda	C. Kowal	1974	20.2	0.001″
Himalia	C. Perrine	1904	15.0	0.05″
Lysithea	S. Nicholson	1938	18.2	0.01″
Elara	C. Perrine	1905	16.6	0.01″
Ananke	S. Nicholson	1951	18.9	0.01″
Carme	S. Nicholson	1938	17.9	0.01″
Pasiphae	P. Melotte	1908	16.9	0.01″
Sinope	S. Nicholson	1914	18.0	0.01″
SATURN				
Pan	Voyager 2	1990	20.0	0.001″
Atlas	R. Terrile	1980	18.0	0.003″
Prometheus	S. Collins	1980	15.8	0.015″
Pandora	S. Collins	1980	16.5	0.015″
Epimetheus	R. Walker	1966	15.7	0.015″
Janus	A. Dollfus	1966	14.5	0.015″
Mimas	W. Herschel	1789	12.9	0.06″
Enceladus	W. Herschel	1789	11.7	0.08″
Tethys	G. Cassini	1684	10.2	0.15″
Telesto	B. Smith	1980	18.7	0.001″
Calypso	B. Smith	1980	19.0	0.001″

This table indicates the discovery dates and the name of the discoverers for most of the satellites in the Solar System. (The 5 faint satellites of Uranus and 12 of Saturn discovered in 1999 and 2000 have not been included.) The magnitudes and apparent sizes of the satellites are also given. The magnitude of a body is its apparent brightness as seen from Earth. It is a logarithmic scale: the brightness decreases by a factor of 100 for an increase of five magnitudes. To set this in context; the brightest stars have magnitudes of about 0, and the faintest stars visible with the naked eye are about magnitude 6. Except for the Moon, none of the satellites in the Solar System is visible with the naked eye, but some twenty-odd of them are visible

Satellite	Discoverer	Year of discovery	Mean apparent magnitude	Mean apparent diameter
SATURN				
Dione	G. Cassini	1684	10.4	0.15″
Helene	P. Laques	1980	18.4	0.001″
Rhea	G. Cassini	1672	9.7	0.25″
Titan	C. Huygens	1655	8.3	0.85″
Hyperion	W. Bond	1848	14.2	0.05″
Iapetus	G. Cassini	1671	11.0	0.25″
Phoebe	W. Pickering	1898	16.5	0.03″
URANUS				
Cordelia	Voyager 2	1986	24	0.002″
Ophelia	Voyager 2	1986	24	0.002″
Bianca	Voyager 2	1986	23	0.003″
Cressida	Voyager 2	1986	22	0.004″
Desdemona	Voyager 2	1986	22	0.005″
Juliet	Voyager 2	1986	22	0.006″
Portia	Voyager 2	1986	21	0.007″
Rosalind	Voyager 2	1986	22	0.003″
Belinda	Voyager 2	1986	22	0.004″
Puck	Voyager 2	1985	20	0.01″
Miranda	G. Kuiper	1948	16.5	0.04″
Ariel	W. Lassell	1851	14.4	0.1″
Umbriel	W. Lassell	1851	15.3	0.1″
Titania	W. Herschel	1787	14.0	0.15″
Oberon	W. Herschel	1787	14.2	0.15″
NEPTUNE				
Naiad	Voyager 2	1989	25	0.0025″
Thalassa	Voyager 2	1989	24	0.004″
Despina	Voyager 2	1989	23	0.0075″
Galatea	Voyager 2	1989	23	0.0075″
Larissa	Voyager 2	1989	22	0.01″
Proteus	Voyager 2	1989	21	0.02″
Triton	W. Lassell	1846	13.6	0.15″
Nereid	G. Kuiper	1949	18.7	0.015″
PLUTO				
Charon	J. Christy	1978	16.8	0.05″

through small amateur telescopes. The date of discovery of each object is a good indicator of the difficulty attending its observation. Finally, the average apparent diameter of each satellite is given. Given that the maximum resolution for amateur telescopes is about 0.5″, most of them remain point sources. By contrast, the Hubble Space Telescope, which photographed the surfaces of Io, Ganymede, and Titan in 1994, and future giant telescopes are capable of partially resolving the surfaces of most of the bodies.

Because astronomy may also be practised without the slightest scientific pretensions. The sight of a sky full of stars is always enchanting, whether seen as one emerges from a prehistoric cave in the Périgord, from the garden of a house in the country, from a deserted beach in the tropics, or from a climbers' refuge high in the mountains. The dome of stars above our heads is, simultaneously, familiar, intimate, and mysterious, as well as perpetually renewed. It prompts us to daydream, places our existence in a far greater context, brings a new perspective to the world, and enriches our thoughts.

Amateur astronomy, whether followed as a scientific hobby or as a new form of exotic tourism, attracts new enthusiasts every year. Could modern-day city-dwellers be feeling a lack of stars? Atmospheric pollution and night-time lighting are preventing people from having daily contact with a vital natural element, one that is rich in poetry and an inexhaustible source of philosophical and metaphysical questions. Could the neon lights of our cities ever replace, in our imagination, the Milky Way and the twinkling of the stars? More prosaically, the general public's current fascination for astronomy may be explained as a cultural phenomenon. The fact that men have walked on the Moon, over a quarter of a century ago, and the conquest of the planets that followed, during the 1980s, opened up new worlds to the West. There was a great temptation to try to visit them – even from a distance, by viewing their light through some form of optical instrument. In fact, for several years now, budding astronomers have had available extremely capable, commercial

equipment, with fine optics and electronics, which are, in effect, small-size replicas of major observatory telescopes. The astronomical performance of these amateur telescopes is, in certain cases – as may be seen from the photographs reproduced here – nothing short of astounding. In fact, although in theory the large telescopes at professional observatories are a thousand times more powerful that those used by amateurs, in practice, the performance of both are more or less placed on an equal footing, or at least brought close together, by turbulence in the Earth's atmosphere, which handicaps large telescopes more than smaller ones. It must be stressed, however, that we are only talking about the observation of the brightest objects,

which are those that interest us here, namely bodies within the Solar System. For observations of faint and distant objects, such as nebulae, galaxies, and quasars, there can be no question that the telescopes used at professional observatories are unparalleled, and their superiority is overwhelming. In broad terms, all optical instruments function on the same principles, whatever their size and their overall power. So the discussions that follow are valid both for the largest telescopes in the world, which are between 3 and 10 metres in diameter, and which weigh several hundred tonnes, as well as for the most modest amateur instrument, which will fit into a suitcase. The key element in a reflecting telescope is its mirror, a disk of polished

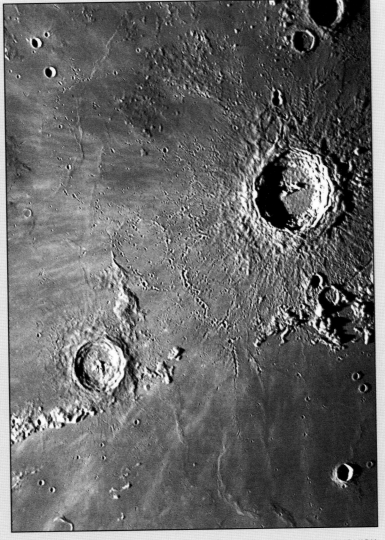

glass whose surface is given a reflective metallic coating. The surface of the mirror is in the form of a parabola, which concentrates the light arriving from celestial objects. The greater the diameter of the mirror, the more light it collects, the sharper and brighter the images that it provides. By way of example, a mirror 6 m in diameter collects one million times more light than the human eye! When such an instrument is linked to a CCD camera, which is more sensitive than the eye and then, in addition, records light from the stars for several hours, it captures more than a billion times more light than the naked eye. In comparison, a good amateur telescope increases the brightness of a celestial object by between 100 and 10 000 times.

So much for brightness. There remains 'magnification', a highly theoretical optical characteristic, about which inexperienced astronomers tend to fantasize, and which is,

■ IN 1980, CCD TECHNOLOGY REVOLUTIONIZED PROFESSIONAL ASTRONOMICAL IMAGERY. IN 2000, AMATEURS ALSO MASTERED THIS METHOD OF OBTAINING IMAGES, AS SHOWN BY THIS MARVELLOUS PORTRAIT OF COPERNICUS, STADIUS AND ERATOSTHENES THAT WAS OBTAINED BY THIERRY LEGAULT WITH HIS 300-MM REFLECTOR AND WIDE-FIELD CCD CAMERA.

■ THIS IMAGE OF SATURN IS ONE OF THE BEST EVER OBTAINED FROM EARTH. THIERRY LEGAULT OBTAINED THIS MAJOR PHOTOGRAPHIC SUCCESS WITH A REFLECTOR JUST 300 MM IN DIAMETER AND A CCD CAMERA. FIFTY INDIVIDUAL HALF-SECOND EXPOSURES WERE MADE TO LIMIT THE EFFECTS OF ATMOSPHERIC TURBULENCE, AND WERE THEN COMBINED ELECTRONICALLY.

however, of little practical interest. The laws of optics (which we will not discuss here), and the effects of atmospheric turbulence limit - at least on Earth – the usable magnification of telescopes, and the more so, the more powerful they are. To summarize, under good seeing conditions, a small instrument, 100 mm in diameter, might be able to magnify the image by between 20 and 200 times. A more powerful instrument, say of 200 mm diameter, might be able to magnify the object by between 50 and 400 times at a maximum. The author, using one of the best instruments, optically, in the world, the famous 1-m telescope at the Pic du Midi, has never exceeded magnifications of 400, 600 or 800 times.

In fact, an experienced observer does not try to obtain a large, indistinct image with a telescope, but rather a bright, sharp image that is precisely suited to their own eyesight. From the optical point of view, this ability is a measurable one, known as resolving power. This is the capacity of an optical instrument to distinguish details on the surface of the body being observed. It is an angular

measurement. To fix our ideas, we may recall that the apparent diameter of the Sun and the Moon is half a degree, or 30' (minutes of arc), or 1800" (seconds of arc). The human eye can resolve details that are about 1' (or 60") apart. The diameters of Venus, Jupiter, and Saturn, which vary as a function of their distances from the Earth, are approximately 30" to 60". Seen with the naked eye, these objects are (obviously) not invisible, but they appear as perfect points of light: the eye is not capable of discerning any details. Mars and Mercury are even smaller, between about 10" to 25". Uranus and Neptune are minute: 4" and 2.5" respectively. Finally, most of the dozens of satellites such as Titan, Triton, Io, Europa, Enceladus, etc., do not even reach 1" in apparent diameter. This figure, 1", is also the maximum resolving power of a telescope 120 mm in diameter. This means that such an instrument would just show these satellites – when they are bright enough – as simple points of light. The planets, the Sun, and the Moon, in contrast have sufficiently large

apparent diameters to allow us to admire, and perceive, a greater or lesser amount of detail.

PLANETS AND AMATEUR ASTRONOMERS

The Solar System is the favourite field for amateur astronomers. Of course, some of them only observe galaxies, while others pass sleepless nights tracking supernovae, but the Solar System always appeals to everyone, whether they are just beginners or highly experienced observers, whether simple skywatchers or cometary experts. The Solar System does, in fact, offer those with small amateur instruments very bright objects (such as the Sun and the Moon, of course), as well as the five planets known since antiquity: Mercury, Venus, Mars, Jupiter, and Saturn.

THE SUN

The Sun has a somewhat equivocal status in the world of amateur astronomy. Because, by definition, it shines only during the daytime, is it really an astronomical object in the true sense? This is uncertain: the magic of

■ CLOSE-UP OF MARE HUMORUM AND THE CRATER BOUILLIAUD. IN THIS PHOTOGRAPH, WORTHY OF A PROFESSIONAL OBSERVATORY, GÉRARD THÉRIN SUCCEEDED IN DETECTING DETAILS THAT ARE JUST A FEW HUNDRED METRES ACROSS. AT THE TOP OF THE PICTURE, ALL THAT REMAINS OF THE ANCIENT CRATER KIES IS ITS ANNULAR RAMPART: THE BOTTOM OF THE CRATER HAS BEEN FLOODED BY THE MOLTEN BASALT FROM MARE HUMORUM.

astronomy, bolstered by its complicated technical rituals, draws its sense of mystery mainly from the fact that it is practised at night. Astronomy, without the night, loses its fascination, its sense, and some of its force.

Yet, rather ironically in a universe where astronomers, above all, have to track down invisible bodies, the Sun is the sole object in the heavens that shines too brightly. Worse than that: observing it without proper precautions can be dangerous, or even fatal, to the eyesight. There is one fundamental piece of advice for all beginners: never aim a telescope or any form of optical instrument at the Sun. With experience, however, and a good knowledge of both what is possible with astronomical equipment, and the procedures to be followed in observing our blinding star, it is certainly possible to study the Sun. Its slowly rotating photosphere and granulation exhibit constantly changing sunspots, some large enough to swallow the Earth several times over, which are the spectacular effects of magnetic storms. Observation of the most beautiful of all solar phenomena, the prominences, which rise from the limb of the Sun, and which may be followed, minute by minute, as they rise, unfortunately requires expensive equipment: either interference filters or a coronograph.

Finally, the Sun's outer atmosphere cannot be observed – and this applies to amateurs as well as to the best-equipped professional astronomers – except for a few minutes every decade, during total eclipses. This extraordinary spectacle is not only astoundingly beautiful, but also extremely rare: there are about three chances per millennium of anyone being able to see an eclipse from any given

■ In yet another image worthy of a professional observatory, taken by Thierry Legault with his 300-mm reflector. The three craters Catharina, Cyrillus, and Theophilus, of very different ages, are easily seen in the smallest amateur telescopes with a magnification of fifty times.

point on Earth. The sole solution for astronomers who want to see one of the most beautiful natural spectacles found in our Solar System, is to travel to the other side of the Earth, if that is where it is visible. Here are a few dates for your diary: Angola in June 2002; Australia in December 2002, Easter Island in July 2010, and the Grand Canyon in February 2012.

The Moon

The Moon is the closest body to us. Even with the naked eye we can clearly make out the great dark plains, ancient basins filled with lava flows, which form the face of the 'Man in the Moon'. Looking at the Earth's satellite with a good pair of binoculars or a small telescope is a revelation, and is often the origin of a passion for astronomy, both among amateurs and professionals. Over the course of a few nights, and with the

succession of lunar phases, the Moon reveals new landscapes, with surprisingly pronounced relief. There are greyish seas that are almost smooth; brilliant highlands, covered in hundreds of craters; and high mountains that cast their inky black shadows far across the ashen-coloured plains. With a more powerful instrument, say one of 200 mm in diameter, one can examine the lunar surface in close-up, and thousands of craters are revealed. At the eyepiece of a telescope it is if one were looking out of a porthole on a spacecraft orbiting the Moon, and one can watch sunrise over the crater Copernicus, for example, which is an imposing ring 90 km in diameter. Towards the centre of the lunar disk, very narrow rilles run across the Triesnecker and Hyginus regions. Elsewhere, experts can search for domes, large hills of volcanic origin, surmounted by crater pits. The amount of extremely fine detail visible on the Moon is amazing: 1″ represents about 1800 m. Because the illumination by the Sun is constantly changing, and the Moon is never at precisely the same angle to the line of sight, the variations in the landscape that may be discovered on this neighbouring world are practically limitless.

Mercury and Venus

These are known as the inferior planets, because they lie between the Sun and the Earth. Like the Moon, Mercury and Venus exhibit a complete cycle of phases. For astronomers, these two planets are difficult to observe: Mercury, the smaller and more distant, never has an apparent diameter that is larger than 4″ to 13″. It is doubtful if anyone has ever truly managed to detect any detail on the surface, and

■ This image shows the sinuous Hadley Rille at the foot of the Apennines. The nearest mountain, Mount Hadley, rises to 5000 m. The Apollo 15 mission landed alongside Hadley Rille. To obtain this CCD image, Thierry Legault used his reflector that is sited just a few tens of kilometres from the centre of Paris.

even managing to detect it in the bright sky close to the Sun is considered quite an achievement by most amateurs.

Venus, on the other hand, is much closer and far brighter, and is thus much easier to observe. In full daylight or in the twilight, it may show a remarkably narrow crescent that is extraordinarily bright. Its apparent diameter varies, as a function of its phase, between 10" and 65". Regrettably, except in rare circumstances, however, when the famous cloud formation like a recumbent 'Y' is detectable, no details are visible with amateur telescopes.

Mars

Entire books have been written about telescopic observations of Mars. Detailed maps of the red planet have been published, beginning about two centuries ago, and these have showed finer and finer detail over time, as discoveries were made and with the progress in astronomical optics. To give a point of reference, we may note that the best telescope in the world for observing the planets, the famous 1-m reflector at the Pic du Midi, allows us to detect the Olympus Mons volcano! The apparent diameter of Mars varies between 3.6" and 26". When, about every two years, the planet exceeds 15", numerous details become accessible to small, amateur telescopes. But the planet Mars is very demanding: its albedo markings are low in contrast, and are detectable only by dedicated and very experienced observers. Nevertheless, perseverance pays: scanning Mars with a telescope of between 200 mm and 1 m in diameter with perfect optics is an unforgettable experience: the evolution of the polar caps, brilliantly white, may be followed from week to week; the major regions, such as Syrtis Major, Sinus Meridiani, and Margaritifer Sinus reveal their variations in colour and in contrast; the high-altitude clouds and the dust-storms may be detected and followed for weeks. Quite apart from the sight that it offers, the red planet also allows amateurs to exercise their talents as artists or photographers, or to participate in international groups that monitor the planets, providing effective support to the professionals. In this way, each space mission, such as Mariner, Viking, or Mars Global Surveyor, is the occasion for amateurs to assist observatories in a thorough study of the red planet before the arrival of the particular spaceprobe.

Jupiter

The giant planet is the favourite target for amateur beginners. Its flattened disk, which measures between 30" and 50" in apparent diameter, clearly shows detail in its bands of clouds. During the course of a single night, the planet, whose rotation period is less than ten hours, literally passes beneath one's eyes. Although a 50-mm refractor, with a resolving power of 2", suffices to show a few vague bands, a 100-mm reflector greatly improves the image, and reveals some of the colours of the jovian clouds. In a 200 or 300-mm reflector, a very experienced observer will be able to pick out dozens of different cloud systems of extraordinary delicacy and variation in tint: salmon pink,

■ AMATEUR ASTRONOMERS HAVE FREQUENTLY REPLACED PROFESSIONALS IN MONITORING THE PLANETS. THIS CCD IMAGE TAKEN BY THIERRY LEGAULT SHOWS FINE DETAIL IN THE ATMOSPHERE OF JUPITER. THE ROTATION OF THE GIANT PLANET IS DISTINCTLY PERCEPTIBLE IN JUST A FEW TENS OF MINUTES OF OBSERVATION.

yellow, cream, ochre, and more. Here again, the study of the giant planet, perhaps lasting over decades, is of very great interest to an amateur who wants to assist the professionals. Finally, even the smallest telescopes allow one to follow the unceasing ballet of the four Galilean satellites. Most of the time they can be seen as tiny luminous points orbiting Jupiter. Frequently, they pass in front of the giant coloured disk, casting a shadow that can be clearly seen as it sweeps across the clouds of hydrogen and helium. Like the Moon and Mars, Jupiter offers a constantly changing spectacle to amateurs who have good equipment, and observe under clear skies free from atmospheric turbulence.

SATURN

Saturn is, without doubt, the jewel of the Solar System. Seen through a telescope, its small, yellowish globe, encircled by its rings, is a beautiful sight, which evokes a magical world or one out of science fiction. Saturn, together with the Moon, is the only body that gives the immediate impression of three-dimensional relief as soon as one looks through the eyepiece.

Although the body of the planet itself shows practically no details, by contrast, the visibility of the wonderful rings changes radically over the course of the years and the decades. The rings lie precisely in the planet's equatorial plane. Because the latter is inclined at 27° to the plane of the ecliptic, the rings appear to undergo a slow, cyclic oscillation over a period of thirty years: the planet's orbital period around the Sun. Between 1987 and 1988 they appeared at maximum opening. Under these optimum conditions, a clear night revealed the Crêpe Ring, as well as the brilliant A and B Rings, separated by the inky black Cassini Division. Such as view was available with a telescope of between 100 and 300 mm in diameter, which also allowed one to observe the subtle interplay of the shadow of the planet cast onto the rings, and of the rings as cast onto the planet. Between August 1995 and February 1996, the rings were seen side-on. They were almost invisible except with telescopes 500 mm to 1 m in diameter. Then, slowly, they reopened until 2002, before starting to close, until 2009, when they will disappear.

URANUS AND NEPTUNE

Because of their distance, the two outermost of the gas giants are invisible to the naked eye. Detectable under good conditions with a telescope of between 100 and 200 mm in diameter, they appear as tiny greenish disks, 4" and 2.5" across, respectively. Need we say that no details are detectable on their dark, vague surfaces?

Pluto is even worse off! This tiny point of light is invisible in small telescopes and requires an instrument with an aperture of about 300 mm to be detected. Only the body's apparent motion over several nights of observation distinguishes it from the stars that are seen in the field of the eyepiece or on the photograph.

Observation of Pluto in fact resembles searching for asteroids, dozens of which may be spotted more or less anywhere, and under the same sort of conditions as the outermost planet in the Solar System.

COMET HUNTING

All that remains is the unpredictable. Dozens of comets are observable, as tiny fuzzy patches, with binoculars or small telescopes. Some, such as Comet Halley, approach sufficiently close to the Earth to develop a long tail, which is spectacular to observe and photograph. The brightest comets are even visible with the naked eye, appearing in the twilit sky above the horizon at sunrise or sunset. An absorbing but very difficult pursuit is hunting for the new comets that are appearing for the first time in the inner Solar System. This is an activity followed by many amateur astronomers, who dream of their name being given to these beautiful visitors to the skies.

BIBLIOGRAPHY AND RECOMMENDED READING

ELEMENTARY AND GENERAL

Beebe, R. (1994), *Jupiter, the Giant Planet*, Smithsonian Institution Press, Washington, D.C. & London

Brunier, S. & Luminet, J.-P. (2000), *Glorious Eclipses*, Cambridge University Press

de la Cotardière, P., ed. (1988), *Larousse Astronomy*, Hamlyn, London

Dodd, R.T. (1986), *Thunderstones and Shooting Stars*, Harvard University Press

Gould, S.J. (1996), *The Mismeasure of Man*, 2nd edn, Norton, New York

Gould, S.J. (1977), *Ever Since Darwin*, Norton, New York

Gould, S.J. (1980), *The Panda's Thumb*, Norton, New York

Gould, S.J. (1989), *Wonderful Life*, Norton, New York

Heidmann, J. (1995), *Extraterrestrial Intelligence*, Cambridge University Press

Hoskin, M., ed. (1997), *Cambridge Illustrated History: Astronomy*, Cambridge University Press

Hunt, G.E. & Moore, P. (1982), *The Planet Venus*, Faber & Faber, London

Illingworth, V. & Clark, J.O.E. (2000), *Facts on File Dictionary of Astronomy*, Facts on File, New York

Lovelock, J. (1989), *The Ages of Gaia*, Bantam Books, London

Maor, E. (2000), *June 8, 2004: Venus in Transit*, Princeton University Press

Moore, P. (1976), *New Guide to the Moon*, Norton, New York

Moore, P. (1982), *The Moon*, Mitchell Beazley, London

Morrison, D. & Samz, J. (1980), *Voyage to Jupiter*, NASA SP 439, NASA, Washington DC

Morrison, D. (1982), *Voyages to Saturn*, NASA SP 451, NASA, Washington DC

Ridpath, I., ed. (1997), *Oxford Dictionary of Astronomy*, Oxford University Press

Sheehan, W. & O'Meara, S.J. (2001), *Mars: The Lure of the Red Planet*, Prometheus

Stern, A. & Mitton, J. (1998), *Pluto and Charon*, Wiley, New York

Whipple, F.L. (1987), *The Mystery of Comets*, Smithsonian Institution Press, Washington DC

Wilson, A. (1987), *Solar System Log*, Janes, New York

INTERMEDIATE

Audouze J. & Israël, G., eds. (1994), *Cambridge Atlas of Astronomy*, 3rd edn, Cambridge University Press

Beatty, J.K., ed. (1999), *The New Solar System*, 4th edn, Cambridge University Press

Burke, J.G. (1986), *Cosmic Debris: Meteorites in History*, University California Press

Carr, M. (1981), *The Surface of Mars*, Yale University Press, New Haven, Conn.

Cattermole, P. (1996), *Planetary Volcanism*, 2nd edn, Wiley, Chichester

Eddy, J.A. (1979), *A New Sun*, NASA SP 402, NASA, Washington DC

Frankel, C. (1996), *Volcanoes of the Solar System*, Cambridge University Press

Greeley, R. (1994), *Planetary Landscapes*, 2nd edn, Chapman & Hall, New York

Greeley, R. & Batson, R.M. (1990), *Planetary Mapping*, Cambridge University Press

Jahns, R.H., ed. (1977), *Skylab explores the Earth*, NASA SP 380, NASA, Washington DC

Kippenhahn, R. (1994), *Discovering the Secrets of the Sun*, Wiley, Chichester

McSween, H.Y. (1987), *Meteorites & Their Parent Planets*, Cambridge University Press

Mark, K. (1987), *Meteorite Craters*, University of Arizona Press, Tucson

Miller, R., Hartmann, W. (1993), *The Grand Tour*, 2nd edn, Workman Publications, New York

Miner, E. (1998), *Uranus: the Planet, Rings and Satellites*, 2nd edn, Wiley, Chichester

Miner, E. & Wessen (2001), R., *Neptune: the Planet, Rings and Satellites*, Springer-Verlag, Heidelberg

Mutch, T.A. & Jones, K.L., eds. (1978), *The Martian Landscape*, NASA SP 425, NASA, Washington DC

Rogers, J.H. (1995), *The Giant Planet Jupiter*, Cambridge University Press

Shea, W.R. (1972), *Galileo's intellectual revolution*, Macmillan, London

Spitzer, C.A., ed. (1980), *Viking Orbiter Views of Mars*, NASA SP 411, NASA, Washington DC

Strom, R.G. (1987), *Mercury: The Elusive Planet*, Cambridge University Press

Yeomans, D.K. (1991), *Comets: A Chronological History ...*, Wiley, New York

ADVANCED OR SPECIALIST

Batson, R.M., Bridges, P.M. & Inge, J.L. (1979), *Atlas of Mars*, NASA, Washington DC

Bergstrahl, J.T., Miner, E.D. & Matthews, M.S. (1991), *Uranus*, University of Arizona Press, Tucson

Bougher, S.W., Hunten, D.M. & Phillips, R.J., eds. (1997) *Venus II*, University of Arizona Press, Tucson

Cruikshank, D.P., ed. (1995) *Neptune and Triton*, University of Arizona Press, Tucson

Davies, M. (1976) *Atlas of Mercury*, NASA, Washington DC

Foukal, P.V., (1990) *Solar Astrophysics*, Wiley, New York

Gehrels, T., ed. (1976) *Jupiter*, University of Arizona Press, Tucson

Gehrels, T., ed. (1984) *Saturn*, University of Arizona Press, Tucson

Gibson, E.G. (1973) *The Quiet Sun*, NASA SP 303, NASA, Washington DC

Greeley, R. & Batson, R. (2001), *Compact NASA Atlas of the Solar System*, Cambridge University Press

Greenberg, R. & Brahic, A., eds. (1984), *Planetary Rings*, University of Arizona Press, Tucson

Hunten, D.M. et al., eds. (1983) *Venus*, University of Arizona Press, Tucson

Kiefer, H.H., et al., eds. (1992) *Mars*, University of Arizona Press, Tucson

Morrison, D., ed. (1982), *Satellites of Jupiter*, University of Arizona Press, Tucson

Schmadel, L.D. (1999), *Dictionary of Minor Planet Names*, 4th edn, Springer-Verlag

Stern, S.A. & Tholen, D.J., eds. (1997), *Pluto and Charon*, University of Arizona Press, Tucson

Vilas, F., Chapman, C.R. & Matthews, M.S. (1988), *Mercury*, University of Arizona Press, Tucson

Wilhelms, D.E., McCauley, J.F. & Trask, N.J. (1987), *The Geologic History of the Moon*, US Geological Survey Professional Paper 1348, USGPO, Washington DC

Wilkening, L.L., ed. (1982), *Comets*, University of Arizona Press, Tucson

INDEX

A

Acidalia Planitia 95
Adams (ring) 200
Adams, John 196
Adonis (asteroid) 114
Africa 52, 54, 60, 64
Ahmad Baba (crater) 36
Airy, George 196
Aldrin, Edwin 74
Algeria 54
Alpha Regio (volcano) 41, 47
Alps 63, 79, 90
Altair 84, 209
Amalthea 201
Amazon 57
Amazonis plain 88
Amazonis (Regio) 93
America 54, 60, 66
Amor 114
Ananke 207
Andes 63, 90
Andros, island of 58
Annapurna (mountain) 54, 66
Antarctica 54, 64, 66
Antares 84
Apennines (mountains) 70, 73, 74, 75, 76, 79
Aphrodite Terra (massif) 46
Apollo (asteroid) 114
Apollo (missions) 76
Apollo 11 (mission) 52
Apollo 11 73, 74
Apollo 12 74
Apollo 13 74
Apollo 15 75, 76
Apollo 16 75
Apollo 17 72, 75, 76, 79, 80, 81
Apollo 18 75
Apollo 19 75
Apollo 20 75
Arago, D. Francois 194
Archaeopteryx (fossil) 63, 66
Ares Vallis (region) 95, 101, 102–103
Argentina 65
Argyre basin 85, 95
Ariel 112, 176, 177
Arizona 61, 115, 129, 209
Armageddon 115
Armstrong, Neil 8, 74
Arsia Mons (volcano) 91, 99, 105, 107
Arthropleura (fossil) 60
Ascraeus Mons (volcano) 91, 99, 105, 107
Asia 52, 54, 64
Asterius Linea 145
Asteroid belt 110, 116
asteroids
    1992 QB$_1$ 114, 210, 212
    1996 TL$_{66}$ 212
Atacama (desert) 64, 65, 75, 82–83
Atlantic Ocean 58, 66
Atlantis 122
Atlas 155, 168, 169

Aulacopleura (fossil) 60
Australia 54, 66, 80–81, 133, 197
Australopithecus afarensis (hominid) 64

B

Bahamas, Archipelago 58
Barnard's Star 187
Beagle 2 (Mars lander) 98
Bean, Alan 74
Belinda 176
Bellatrix 19
Beppi-Colombo (probe) 36
Beta Pictoris 22
Beta Regio (massif) 46
Bianca 176
Big Bang 18
Bolivia 203
Bouvard (astronomer) 194
Brachiosaurus (fossil) 60

C

California 129, 197
Callisto 36, 72, 79, 142, 143, 145, 148, 149, 164, 169, 211
Caloris Basin 35, 36
Calypso 169
Canopus 19
Carme 201
Cassini (probe) 155, 165, 201
Cassini Division 150–151, 155, 159
Cassini-Huygens Mission 141, 165, 168, 171
Castalia 112
Cerberus (region) 93
Ceres 108–109, 112, 114
Cernan, Eugene 79
Charon 72, 112, 201, 202, 205, 207, 208–213
Chiron 114
Chicxulub, Mexico 63
Chile 64, 65, 106, 133, 203, 210
Chryse (plain) 86, 89, 93, 103, 105
Chryse Planitia 84
Cleopatra (crater) 46
Colorado (canyon) 61
Comte, Auguste 13, 194
Conrad, Charles 74
Copernicus (crater) 79
Coprates Chasma (region) 90–91, 92
Cordelia 174, 176
Cornwall 66
Corot (satellite) 23
Cressida 176
Cuba 58

D

Dactyl 110, 111, 112, 113
Danilova (crater) 44
Darwin, Charles 59
Darwin (ESA project) 23
Daulaghiri (mountain) 54, 200
Dead Sea 53
Deimos 84, 93, 100, 104–107, 110, 126, 180
Deneb 19, 84, 209

Denitsa Regio 40
Descartes (crater) 72, 75
Desdemona 176
Dione 112, 154, 168, 169, 170, 171, 211
Diplodocus (fossil) 60
Discovery (Space Shuttle) 59
Doerfel, Montes 76
Duke, Charles 72, 75

E

E Ring (Saturn) 171
Earth 14–15, 16, 18, 20, 21, 22, 23, 28, 29, 34, 36, 37, 38, 41, 42, 43, 44, 48–67, 48–49, 50, 52, 57, 62, 63, 67, 68–69, 70, 71, 72, 73, 74, 75, 76, 79, 80, 80–81, 81, 82–83, 86, 87, 89, 93, 95, 97, 98, 106, 107, 110, 105, 112, 114, 116–117, 118, 119, 121, 122, 136, 137, 139, 143, 152, 154, 155, 157, 160–161, 163, 164, 165, 167, 168, 171, 174, 176, 177, 178–179, 182, 182–183, 184, 185, 186, 187, 194, 196, 197, 198, 199, 200–201, 203, 205, 206, 209, 211, 212
Easter Island 67
École polytechnique 194
Egypt 53
Eleuthera (island) 58
Elysium (region) 93
Enceladus 53, 158, 166–171, 180
Encke 187, 191
England 196
Epimetheus 168, 169
Epsilon (ring) 177
Eridania (region) 89
Eros 113, 114, 115
Eta Carinae 20–21
Eugenia 112
Eurasia (continent) 54, 60
Europa 52, 66
Europa (satellite) 36, 53, 72, 116–117, 136, 142–149, 164, 165, 171, 205, 211
Europa Orbiter 147
European Space Agency (ESA) 165, 191
Eusthenopteron (fossil) 60, 64
Everest 106
Exumas (island) 58

F

Flanders 66
Florida 58
Fra Mauro (crater) 74
France 133
Fresco (at Ouan Rechla) 54

G

Gaia 64
Galaxy 14–15, 16, 18, 19, 20, 22, 23, 28, 30, 187, 200–201
Galileo 130, 196
Galileo (probe) 110, 111, 112, 113, 122, 124, 125, 131, 133, 139, 140, 141, 145, 146, 147, 201
Galileo Regio 147, 148

Galle, Johann 196
Ganges 51
Ganymede 22, 36, 72, 136, 142, 143, 145,
    146, 147, 148, 149, 162, 164, 205, 211
Gaspra 108–115, 122, 126, 180
Geographos (asteroid) 114
Giotto (probe) 180, 184, 191
Gobi Desert 66
Gondwana (continent) 54, 60
Great Dark Spot 194, 198
Great Red Spot 119, 120, 121, 123, 133, 198
Guadalquivir (valley) 66
Gula Mons (volcano) 46–47
Gulf Stream 66

H

Hadley Mons 76
Haemus Mons 136
Hale Telescope 184
Hale–Bopp (Comet) 188–191
Hall (crater) 106
Halley (Comet) 178–191
Halley, Edmond 183
Hawaiian islands 54, 59, 135, 139
Helena 169
Hellas basin 87, 89, 95, 101
Hermes (asteroid) 114
Herschel, William 174, 194
Hidalgo 210
Himalayas 50, 54, 63, 64, 65, 66, 89
Hoggar 93
Homo 66
Homo erectus (hominid) 66, 100
Homo habilis (hominid) 66
Homo sapiens 64, 66, 100
Homo sapiens sapiens 64, 66
Howard-Koomen (Comet) 191
Hubble Space Telescope 22, 23, 98, 121,
    122, 130, 131, 132, 133, 139, 140, 154,
    157, 165, 176, 208, 211, 212
Humboldt Current 64
Huygens (module) 165
Huygens Mons 76
Hyakutake (Comet) 187–188
Hydra 17
Hyginus Rille (craterlets) 68–69
Hyperion 169, 171

I

Icarus (asteroid) 114, 115
Ida 110, 111, 112, 113, 122
Ikeya-Seki (Comet) 187
India 50, 54
Indian Ocean 64, 66
Iniki (cyclone) 59
Internet 131
Io 36, 53, 72, 116–117, 121, 126–127,
    134–141, 142, 143, 144, 145, 146, 148,
    164, 171, 176, 205, 211
Irwin, James 73, 74, 75
Ishtar Terra (high plateau) 46, 47
Israel 53
Ithaca Chasma (canyon) 169

J

Jacquard, Albert 59
Janus 157, 168, 169

Japetus 169, 211
Juliet 176
Juno 16
Jupiter 21, 22, 36, 37, 72, 110, 112,
    116–127, 128–145, 146, 148, 149, 152,
    154, 155, 159, 162, 164, 172, 173, 174,
    176, 177, 185, 187, 194, 196, 198, 199,
    200, 201, 206, 211, 212

K

Kali Gandakhi (river) 51
Kamtchatka Peninsula 50
Kepler (crater) 79
Kepler (satellite) 23
Kohoutek (Comet) 170
Krakatoa (volcano) 139

L

Laguna Verde (region) 64, 65
Lalande, Joseph 196
Laplace, Pierre Simon 194
Lascaux (cave) 64
Laurasia (continent) 54, 60
Le Verrier (ring) 200
Le Verrier, Urbain 194, 196
Leh (region) 66
Leibniz Mons 76
Levy David (astronomer) 129
Loki (volcano) 135, 137, 139
Lunae Planum (region) 93

M

M 42 (Orion Nebula) 22
Maasaw Patera (volcano) 135
Machhapuchhare (mountain) 50, 51, 54, 66
Madagascar 54
Magellan (probe) 41, 44, 45, 46, 46–47, 47, 59
Maja Vallis 93
Mali 54
Mare Caloris (basin) 34
Mare Crisium 79
Mare Imbrium 79
Mare Serenitatis 74, 78, 79, 80
Mare Tranquilitatis 74
Mariner 10 (probe) 36, 44
Mars 21, 22, 23, 37, 43, 46, 50, 53, 64, 76,
    81, 82–103, 84, 85, 86, 87, 89, 92, 98,
    100, 101, 102, 103, 104, 105, 106, 107,
    110, 112, 114, 126, 130, 136, 143, 145,
    148, 171, 174, 176, 180
Mars 1996 (probe) 96
Mars Climate Orbiter (probe) 95
Mars Express (mission) 98
Mars Global Surveyor (MGS) 96, 97
Mars Observer (probe) 96
Mars Odyssey (probe) 97
Mars Pathfinder (probe) 95, 96, 97, 100,
    101, 102
Mars Polar Lander (probe) 95, 96
Mars Rover 2003 (probe) 97
Mathilde 113
Mattingly, Thomas 72
Maxwell Montes (massif) 44, 45, 46
Méchain, Pierre 194
Meganeura (fossil) 60
Megarachne 60
Megazostrodon (fossil) 63

Mercury 21, 22, 26, 28, 30, 32–37, 40, 42,
    44, 46, 53, 72, 76, 79, 104, 106, 107,
    110, 114, 126, 130, 136, 148, 162–164,
    169, 174, 176, 182, 211, 212
Messenger (probe) 36
Meteor Crater 115
Mexico 63
Mexico, Gulf of 66
Mie (crater) 93
Mimas 157, 158, 159, 168, 169, 171, 180
Minos linea 145
Milky Way 16, 17, 19, 20 , 22, 23, 28, 37,
    57, 84, 86, 110, 178–179, 180, 208–209
Miranda 176–177
Mitchell, Edgar 74
Modi Khola (river) 51
Modocia 60
Montana 64
Moon 22, 28, 32–33, 36, 37, 46, 52, 53, 62,
    68–81, 84, 86, 90, 104, 106, 110,
    116–117, 121, 126, 134, 135, 136, 139,
    142, 143, 145, 148, 162, 164, 168, 169,
    174, 176, 203, 205, 211
Movile Cave, Romania 103

N

NASA 86, 141, 147, 165, 196, 197, 209
Nassau 58
Near (probe) 113, 114, 115
Nepal 50
Neptune 21, 22, 72, 118, 122, 154, 159,
    165, 172–173, 174, 176, 177, 182, 184,
    192–201, 202, 203, 205, 206, 207, 211,
    212
Nereid 199, 200
Nevado Ojos del Salado (volcano) 64, 65
Niger 54
Nile 53
Niobe Planitia (lava flow) 47
Noachis (region) 89
Noctis Labyrinthus 99

O

Oberon 176, 177
Observatories
    Berlin 196
    Calar Alto, Andalusia 131
    Haute-Provence 22
    Kitt Peak, Arizona 129
    La Silla, Chile 172
    Mauna Kea, Hawaii 112, 130–133, 185
    Mount Palomar 129, 184
    Paris 186
    Sacramento Peak, New Mexico 30
    Siding Spring 128
Oceanus Borealis (region) 95, 96, 103
Oceanus Procellarum 74, 79
Olenellus (fossil) 60, 66
Olenellus nevadensis 60
Olympus Mons (volcano) 90, 91, 93, 94, 95,
    105, 107
Oort, Jan Hendrik 173
Oort Cloud 187, 212
Ophelia 176, 201
Ophir Point 92
Orion Nebula 22, 23
Ouan Rechla (basin) 50

Ouan Rechla, Lake at 54, 64
Ovda Regio 44

P

Pacific Ocean 50, 54, 59, 67, 80–81
Pallas 112
Pamir (mountains) 54, 65
Panama Isthmus 66
Pandora 158, 168, 169
Pangaia (continent) 54, 60
Pasiphae 207
Pavonis Mons (volcano) 91, 99, 105, 107
Pegasi 51 (exoplanet) 22
Pele (volcano) 135, 137, 138
Périgord 66
Phobos 22, 84, 90, 100, 104–107, 110, 112, 126, 180
Phoebe 169, 207
Phoebe Regio 42
Pikaia gracilens (fossil) 59, 60
Pillan Patera 140, 141
Pinatubo (volcano) 139
Pioneer Venus (probe) 44
Pluto 22, 36, 72, 114, 141, 148, 162, 164, 194, 198, 201, 202, 205, 206, 207, 208–213, 194
Pole Star (Polaris) 19
Portia 176
Prometheus (satellite) 158, 168, 169, 201
Prometheus (volcano) 135, 139, 141
Proteus 198
Provence 64
Proxima Centauri 187, 211, 212
Pteranodon (fossil) 60
Puck 176
Puerto Rico 58

Q

Quetzalcoatlus (fossil) 60

R

Ra Patera (volcano) 135, 137
Ravi Vallis (region) 100
Red Sea 53
Rhea 167, 168, 169, 170, 176, 211
Rigel 19
Roche (crater) 106
Roche Limit 107, 130, 159, 207
Rosalind 176
Romania 103

S

Sacajawea Patera (volcano) 47
Sagittarius 18
Sahara Desert 50, 54, 55, 66, 82–83, 86
Saint-George (crater) 74
Saiph 19
Satellite system 199
Saturn 21, 22, 36, 37, 53, 72, 114, 118, 121, 150–159, 160–161, 162, 164, 165, 166, 167, 168, 169, 171, 172–173, 174, 176, 177, 180, 182, 184, 185, 187, 194, 196, 198, 200, 201, 206, 207, 210, 211

Saudi Arabia 53
Schmitt, Harrison 75, 79, 80
Schwassmann-Wachmann (comet) 187
Scott, David 73, 74, 75
Shepard, Alan 74
Shoemaker, Eugene and Carolyn (astronomers) 129
Shoemaker-Levy (comet) 128–133, 185, 191
Sif Mons (volcano) 46–47
Sinai (massif) 53
Sinope 207
Sinus Medii (region) 68–69
Sinus Meridiani (region) 101,103
Sirius 22
Skylab 184
Soho (satellite) 29, 31
Solar System 18, 21, 22, 23, 26, 30, 32–33, 34, 36, 37, 46, 47, 48–49, 53, 54, 57, 62, 70, 72, 75, 79, 80, 81, 84, 89, 91, 93, 93, 102, 107, 108–109, 110, 112, 113, 116–117, 118, 120, 121, 123, 126, 127, 129, 133, 134, 135, 136, 137, 139, 140, 141, 143, 145, 148, 149, 150–151, 152, 153, 154, 159, 160–161, 162, 164, 165, 167, 168, 169, 171, 172–173, 174, 175, 176, 177, 180, 182, 182–183, 187, 189, 191, 192–193, 194, 196, 198, 199, 200–201, 201, 202, 203, 205, 206, 207, 208, 209, 211, 212
South America 64, 67
Spain 133, 197
Spitzbergen 66
Spot (satellite) 65
Stars 14–23, 14–15, 20
Stickney (crater) 106, 107
Strindberg (crater) 36
Suez Canal 53
Sun 16, 19, 20, 21, 22, 23, 24–31, 32–33, 34, 36, 37, 38–39, 40, 42, 43, 44, 47, 50, 53, 59, 67, 70, 71, 86, 87, 93, 103, 110, 112, 118, 122,126, 127, 129, 130, 131, 133, 135, 143, 152, 154, 155, 160–161, 162, 164, 165, 167, 168, 169, 171, 174, 176, 177, 178–179, 180, 182, 183, 184, 185, 187, 189, 194, 198, 199, 203, 205, 206, 207, 209, 211, 212
Surveyor 3 (probe) 74
Syrtis Major (region) 87, 89, 98

T

Takla Makan (region) 66, 75, 93
Tambora (volcano) 139
Taurus (constellation) 209
Taurus-Littrow valley 74, 79, 80
Telesto 169
Tethys 112, 150–151, 154, 168, 169, 171, 211
Tharsis (plateau) 92, 98, 99, 105
Titan 22, 36, 53, 72, 136, 148, 160–165, 166, 168, 169, 170, 171, 174, 177, 199, 205, 206, 211
Titan-Centaur 122
Titania 176

Tithonium Chasma (region) 92
Tiu Vallis 95
Tombaugh, Clyde 209, 211
Toutatis (asteroid) 112, 115
Triceratops (fossil) 63, 66
Triesnecker (crater) 68–69
Triton 22, 36, 72, 136, 141 164, 165, 198, 199, 200, 201, 202–207, 211
Tycho (crater) 79
Tyrannosaurus rex (fossil) 63

U

Ultrasaurus (fossil) 60
Umbriel 112, 176
Universe 16, 18, 19, 23, 48–49, 57, 72, 78, 177, 191
Uranus 21, 22, 114, 118, 122, 154, 159, 172–177, 187, 194, 196, 198, 200, 201, 206, 210, 211, 212
Utopia plain 84, 86, 89, 93, 103, 105
Utopia Planitia 89, 95, 96–97, 98

V

Valles Marineris (canyons) 90–91, 92, 97, 98, 99, 105
Vastitas Borealis (region) 92, 95
Vega 84, 209
Vega 1 (probe) 44
Vega 2 (probe) 44
Venera (probes) 45, 46
Venera 9 (probe) 45
Venera 10 (probe) 45
Venera 13 (probe) 40, 42–43, 45
Venera 14 (probe) 40, 42–43, 45
Venus 21, 22, 23, 37, 38–47, 38–39, 40, 41, 42–43, 45, 46–47, 53, 59, 64, 72, 76, 90, 107, 112, 125, 136, 162, 164, 171, 174, 177, 182
Vesta 112
Viking (probes) 86, 95, 101, 102, 106
Viking 1 (probe) 46, 89, 95
Viking 2 (probe) 46, 87, 89, 93, 95, 96–97, 106, 107
Voyager (missions) 171
Voyager 2 (mission) 196
Voyager (probes) 122, 137, 145, 158, 164, 176, 196
Voyager 1 (probe) 118, 121, 122, 138, 139, 155, 168
Voyager 2 (probe) 121, 122, 144, 155, 159, 162, 172–173, 174, 175, 176, 192–193, 196–197, 198, 199, 200–201, 202

W

West (Comet) 191

Y

Young, John 72, 75

PHOTOGRAPHIC CREDITS

INTRODUCTION: THE STORY OF THE PLANETS
p.6–7: S. Brunier / Ciel & Espace
p.9: S. Brunier / Ciel & Espace
p.11: NASA / Ciel & Espace
p.12: S. Brunier / Ciel & Espace

A STAR LOST IN INFINITE SPACE
p.14–15: D. Malin / AAO / Ciel & Espace
p.16: D. Malin / AAO / Ciel & Espace
p.17: D. Malin / AAO / Ciel & Espace
p.18: *top*, A. Fujii / Ciel & Espace
p.18: *centre*, D. Malin / ROE / Ciel & Espace
p.18: *bottom*, D. Malin / ROE / Ciel & Espace
p.19: D. Malin / AAO / Ciel & Espace
p.20–21: D. Malin / AAO / Ciel & Espace
p.22: *top*, NASA / ESA / Ciel & Espace
p.22: *bottom*, NASA / ESA / Ciel & Espace
p.23: NASA / ESA / Ciel & Espace

THE SUN: OUR STAR
p.24–25: BBSO / O. Hodasava / Ciel & Espace
p.26: NOAA / Ciel & Espace
p.27: Trace / NASA / Ciel & Espace
p.28: *top*, NOAA / O. Hodasava / Ciel & Espace
p.28: *bottom*, ISAS / Ciel & Espace
p.29: NOAA / Ciel & Espace
p.30: Sacramento Peak / NOAA / Ciel & Espace
p.31: NASA / ESA / Ciel & Espace

MERCURY: BAKED BY THE HEAT OF THE SUN
p.32–33: NASA / O. Hodasava / Ciel & Espace
p.34: NASA / O. Hodasava / Ciel & Espace
p.35: NASA / O. Hodasava / Ciel & Espace
p.36: NASA / O. Hodasava / Ciel & Espace
p.37: NASA / O. Hodasava / Ciel & Espace

VENUS: A VISION OF HELL
p.38–39: NASA / Ciel & Espace
p.40: JPL / NASA / O. Hodasava / Ciel & Espace
p.41: JPL / NASA / O. Hodasava / Ciel & Espace
p.42–43: IKI / O. Hodasava / Ciel & Espace
p.44: JPL / NASA / O. Hodasava / Ciel & Espace
p.45: JPL / NASA / Ciel & Espace
p.46–47: JPL / NASA / Ciel & Espace

THE EARTH: THE STORY OF A LIVING PLANET
p.48–49: NASA / Ciel & Espace
p.50: NASA / Ciel & Espace
p.51: S. Brunier / Ciel & Espace
p.52: NASA / Ciel & Espace
p.53: NASA / Ciel & Espace
p.54: S. Brunier / Ciel & Espace
p.55: NASA / Ciel & Espace
p.56: NASA / Ciel & Espace
p.57: S. Brunier / Ciel & Espace
p.58: NASA / Ciel & Espace
p.59: NASA / Ciel & Espace
p.60: S. Brunier / Ciel & Espace
p.61: S. Brunier / Ciel & Espace
p.62: NASA / Ciel & Espace
p.63: NASA / Ciel & Espace
p.64: S. Brunier / Ciel & Espace
p.65: CNES / Spotimage
p.66: S. Brunier / Ciel & Espace
p.67: NASA / Ciel & Espace

THE MOON: SETTING FOOT ON ANOTHER WORLD
p.68–69: NASA / Ciel & Espace
p.70: NASA / Ciel & Espace
p.71: NASA / Ciel & Espace
p.72: NASA / Ciel & Espace
p.73: NASA / Ciel & Espace
p.74: NASA / Ciel & Espace
p.75: NASA / Ciel & Espace
p.76: NASA / Ciel & Espace
p.77: NASA / Ciel & Espace
p.78: NASA / Ciel & Espace
p.79: NASA / Ciel & Espace
p.80–81: NASA / Ciel & Espace

MARS: A TRIP TO THE DESERT PLANET
p.82–83: NASA / Ciel & Espace
p.84: NASA / Ciel & Espace
p.85: NASA / Ciel & Espace
p.86: USGS / NASA / Ciel & Espace
p.87: USGS / NASA / Ciel & Espace
p.88: USGS / NASA / Ciel & Espace
p.89: *top*, USGS / NASA / Ciel & Espace
p.89: *bottom*, NASA /Ciel & Espace
p.90–91: USGS / NASA / Ciel & Espace
p.92: *top*, USGS / NASA / Ciel & Espace
p.92: *bottom*, NASA / Ciel & Espace
p.93: USGS / NASA / Ciel & Espace
p.94: USGS / NASA / Ciel & Espace

p.95: USGS / NASA / Ciel & Espace
p.96–97: *top*, JPL / NASA / Ciel & Espace
p.96: *bottom*, JPL / NASA / Ciel & Espace
p.97: *bottom*, JPL / NASA / Ciel & Espace
p.98: *top*, JPL / NASA / Ciel & Espace
p.98: *bottom*, NASA / ESA / Ciel & Espace
p.99: USGS / NASA / Ciel & Espace
p.100: NASA / Ciel & Espace
p.101: NASA / Ciel & Espace
p.102–103: NASA / Ciel & Espace

PHOBOS AND DEIMOS: PEBBLES IN THE SKY
p.104: NASA / O. Hodasava / Ciel & Espace
p.105: NASA / O. Hodasava / Ciel & Espace
p.106: NASA / Ciel & Espace
p.107: NASA / O. Hodasava / Ciel & Espace

GASPRA: OUR FIRST ASTEROID
p.108–109: JPL / NASA / Ciel & Espace
p.110: JPL / O. Hodasava / Ciel & Espace
p.111: JPL / Ciel & Espace
p.112: JPL / NAIC / NSF / Ciel & Espace
p.113: NASA / Ciel & Espace
p.114: NASA / Ciel & Espace
p.115: NASA / Ciel & Espace

JUPITER: THE PLANET OF STORMS
p.116–117: JPL / NASA / Ciel & Espace
p.118: JPL / NASA / Ciel & Espace
p.119: JPL / NASA / Ciel & Espace
p.120: JPL / NASA / Ciel & Espace
p.121: *top*, NASA / ESA / Ciel & Espace
p.121: *centre and bottom*, JPL / NASA / Ciel & Espace
p.122: NASA / Ciel & Espace
p.123: JPL / NASA / Ciel & Espace
p.124: NASA / Ciel & Espace
p.125: NASA / Ciel & Espace
p.126–127: JPL / NASA / Ciel & Espace

SHOEMAKER-LEVY: TIMETABLE TO COLLISION
p.128: AAT / Ciel & Espace
p.129: MPI / Ciel & Espace
p.130–131: NASA / Ciel & Espace
p.131: *bottom*, MPI / Ciel & Espace
p.132: NASA / Ciel & Espace
p.133: NASA / Ciel & Espace

IO: THE VOLCANO PLANET
p.134: JPL / NASA / Ciel & Espace
p.135: JPL / NASA / Ciel & Espace
p.136: JPL / NASA / Ciel & Espace
p.137: USGS / NASA / Ciel & Espace
p.138: JPL / NASA / Ciel & Espace
p.139: JPL / NASA / Ciel & Espace
p.140: NASA / Ciel & Espace
p.141: NASA / Ciel & Espace

EUROPA: THE HIDDEN OCEAN
p.142: USGS / NASA / Ciel & Espace
p.143: JPL / NASA / Ciel & Espace
p.144: USGS / NASA / Ciel & Espace
p.145: NASA / Ciel & Espace
p.146: NASA / Ciel & Espace
p.147: NASA / Ciel & Espace
p.148: JPL / NASA / Ciel & Espace
p.149: JPL / NASA / Ciel & Espace

SATURN: THE LORD OF THE RINGS
p.150–151: JPL / NASA / Ciel & Espace
p.152: JPL / NASA / Ciel & Espace
p.153: JPL / NASA / Ciel & Espace
p.155: NASA / Ciel & Espace
p.156: JPL / NASA / Ciel & Espace
p.157: NASA / ESA / Ciel & Espace
p.158: JPL / NASA / Ciel & Espace
p.159: JPL / NASA / Ciel & Espace

TITAN: AN EARTH IN HIBERNATION
p.160–161: JPL / NASA / Ciel & Espace
p.162: JPL / NASA / Ciel & Espace
p.163: JPL / NASA / Ciel & Espace
p.164: JPL / NASA / Ciel & Espace
p.165: NASA / Ciel & Espace

ENCELADUS AND THE WORLDS OF ICE
p.166: JPL / NASA / Ciel & Espace
p.167: JPL / NASA / Ciel & Espace
p.168: JPL / NASA / Ciel & Espace
p.169: JPL / NASA / Ciel & Espace
p.170: JPL / NASA / Ciel & Espace
p.171: JPL / NASA / Ciel & Espace

URANUS: A RECUMBENT GIANT
p.172–173: JPL / NASA / Ciel & Espace
p.174: JPL / NASA / Ciel & Espace

p. 175: JPL / NASA / Ciel & Espace
p. 176: NASA / Ciel & Espace
p. 177: *top*, JPL / NASA / Ciel & Espace
p. 177: *bottom*, USGS / NASA / Ciel & Espace

HALLEY: THE GREAT TRAVELLER
p. 178–179: A. Fujii / Ciel & Espace
p. 180: ESA / Ciel & Espace
p. 181: D. Malin / AAO / Ciel & Espace
p. 182–183: D. Malin / AAO / Ciel & Espace
p. 184: NASA / Ciel & Espace
p. 185: D. Jewitt / J. Luu / U.H. / Ciel & Espace
p. 186: A. Fujii / Ciel & Espace
p. 187: S. Brunier / Ciel & Espace
p. 188: S. Anglaret / Ciel & Espace
p. 189: J. Lodrigues / Ciel & Espace
p. 190: A. Fujii / Ciel & Espace
p. 191: A. Fujii / Ciel & Espace

NEPTUNE: THE GREAT BLUE SEA
p. 192–193: JPL / NASA / Ciel & Espace
p. 194: JPL / NASA / Ciel & Espace
p. 195: JPL / NASA / Ciel & Espace
p. 196: JPL / NASA / Ciel & Espace
p. 197: JPL / NASA / Ciel & Espace
p. 198: JPL / NASA /Ciel & Espace
p. 199: JPL / NASA / Ciel & Espace
p. 200–201: JPL / NASA / Ciel & Espace

TRITON: VOLCANOES OF ICE
p. 202: USGS / NASA / Ciel & Espace
p. 203: USGS / NASA / Ciel & Espace
p. 204: NASA / O. Hodasava / Ciel & Espace
p. 205: USGS / NASA / Ciel & Espace
p. 206: USGS / NASA / Ciel & Espace
p. 207: JPL / NASA / Ciel & Espace

PLUTO AND CHARON: PLANETS IN LIMBO
p. 208: NASA / ESA / Ciel & Espace
p. 209: C. Tombaugh / Lowell Obs / Ciel & Espace
p. 210: ESO / Ciel & Espace
p. 211: NASA / ESA / Ciel & Espace
p. 212: NASA / Ciel & Espace
p. 213: ESO / Ciel & Espace

POSITIONS AND MOTIONS IN THE SOLAR SYSTEM
p. 214: P. Roth / O. Hodasava / Ciel & Espace
p. 215: S. Brunier / O. Hodasava / Ciel & Espace
p. 216: S. Brunier / O. Hodasava / Ciel & Espace
p. 220: S. Brunier / Ciel & Espace
p. 221: A. Fujii / Ciel & Espace

TELESCOPES AND SPACEPROBES
p. 226: M. Robinson / UH / USGS / Ciel & Espace
p. 227: S. Brunier / Ciel & Espace
p. 230: NASA / Ciel & Espace
p. 232: NASA / Ciel & Espace
p. 233: USGS / Ciel & Espace

AMATEUR OBSERVATION OF THE PLANETS
p. 234: G. Thérin / Ciel & Espace
p. 235: C. Arsidi / Ciel & Espace
p. 238: Th. Legault
p. 239: Th. Legault
p. 240: G. Thérin / Ciel & Espace
p. 241: Th. Legault
p. 242: Th. Legault
p. 243: Th. Legault

ABBREVIATIONS:
AAO   Anglo-Australian Observatory
AAT   Anglo-Australian Telescope
BBSO Big Bear Solar Observatory, USA
CNES Centre Nationale de Recherches Spatiales, France
ESA   European Space Agency
ESO   European Southern Observatory
IKI   Space Research Institute, Russian Academy of Sciences, Moscow
ISAS Institute of Space and Astronautical Science, Japan
JPL   Jet Propulsion Laboratory, Pasadena, USA
MPI   Max-Planck Institut, Garching, Germany
NAIC National Astronomy and Ionosphere Center, USA
NASA National Aeronautics and Space Administration, USA
NOAA National Oceanic and Atmospheric Administration, USA
NSF   National Science Foundation, USA
ROE   Royal Observatory, Edinburgh, UK
UH   University of Hawaii
USGS United States Geological Survey